THE LONG HISTORY OF THE FUTURE

Also available in the Bloomsbury Sigma series:

Sex on Earth by Jules Howard
Spirals in Time by Helen Scales
A is for Arsenic by Kathryn Harkup
Suspicious Minds by Rob Brotherton
Herding Hemingway's Cats by Kat Arney
The Tyrannosaur Chronicles by David Hone
Soccermatics by David Sumpter
Goldilocks and the Water Bears by Louisa Preston
Science and the City by Laurie Winkless
Built on Bones by Brenna Hassett
The Planet Factory by Elizabeth Tasker
Catching Stardust by Natalie Starkey
Nodding Off by Alice Gregory
Turned On by Kate Devlin
Borrowed Time by Sue Armstrong
The Vinyl Frontier by Jonathan Scott
Clearing the Air by Tim Smedley
Superheavy by Kit Chapman
The Contact Paradox by Keith Cooper
Life Changing by Helen Pilcher
Kindred by Rebecca Wragg Sykes
Our Only Home by His Holiness The Dalai Lama
First Light by Emma Chapman
Ouch! by Margee Kerr & Linda Rodriguez McRobbie
Models of the Mind by Grace Lindsay
The Brilliant Abyss by Helen Scales
Overloaded by Ginny Smith
Handmade by Anna Ploszajski
Beasts Before Us by Elsa Panciroli
Our Biggest Experiment by Alice Bell
Worlds in Shadow by Patrick Nunn
Aesop's Animals by Jo Wimpenny
Fire and Ice by Natalie Starkey
Sticky by Laurie Winkless
Wilder by Millie Kerr
The Tomb of the Milli Monga by Samuel Turvey
Superspy Science by Kathryn Harkup
Warming Up by Madeleine Orr
Into the Groove by Jonathan Scott

THE LONG HISTORY OF THE FUTURE

WHY TOMORROW'S TECHNOLOGY STILL ISN'T HERE

Nicole Kobie

B L O O M S B U R Y S I G M A
LONDON • OXFORD • NEW YORK • NEW DELHI • SYDNEY

BLOOMSBURY SIGMA
Bloomsbury Publishing Plc
50 Bedford Square, London, WC1B 3DP, UK
29 Earlsfort Terrace, Dublin 2, Ireland

BLOOMSBURY, BLOOMSBURY SIGMA and the Bloomsbury Sigma logo are trademarks of Bloomsbury Publishing Plc

First published in the United Kingdom in 2024

A catalogue record for this book is available from the British Library

Library of Congress Cataloguing-in-Publication data has been applied for

ISBN: HB: 978-1-3994-0310-8; eBook: 978-1-3994-0311-5

2 4 6 8 10 9 7 5 3 1

Typeset in Bembo Std by Deanta Global Publishing Services, Chennai, India
Printed and bound in Great Britain by CPI Group (UK) Ltd, Croydon CR0 4YY

To Michael, for everything, and to Eliza,
because it’s her future

Contents

Introduction 9

Chapter 1: Driverless Cars 19

Chapter 2: Artificial Intelligence 65

Chapter 3: Robots 109

Chapter 4: Augmented Reality 147

Chapter 5: Cyborgs and Brain–Computer Interfaces 181

Chapter 6: Flying Cars 217

Chapter 7: Hyperloop 249

Chapter 8: Smart Cities 283

Chapter 9: Who Builds the Future? 323

Acknowledgements 335

Bibliography 337

Index 359

INTRODUCTION

It's been said that the best way to predict the future is to create it.*

That may be true, but have you ever tried building anything?

Driverless cars, flying ones, superintelligent artificial intelligence (AI), humanoid robots, cyborgs – clever scientists, genius engineers and self-taught polymaths have been trying to build these visions of the future for decades or longer. Yet, here we are: driving licences for road-bound cars in our pockets, domestic chores still done by our own meaty hands, our thinking limited by our brains rather than extended by machines.

Our failure to build these future technologies has nothing to do with a lack of effort. The brightest minds of multiple generations have spent lifetimes developing these ideas at top universities, famous labs, Silicon Valley campuses and startup garages – not to mention the eccentric inventors who go it on their own.

And we've had success. Neural networks aren't new; they've been in the works since the 1940s. Though our commutes aren't airborne, the first flying cars were licensed to fly in the 1950s. The first augmented reality headset was plonked on a student's head in the 1960s. Today, most people have smartphones, the internet is almost ubiquitous and space

* Variations on this quote have been attributed to many different people. It's pasted over Abraham Lincoln's face on plenty of social media memes, but that's certainly misquoted. Management consultant Peter Drucker did say it, though Alan Kay of the famous PARC research centre in Palo Alto may have said a version first. In turn, Kay may have been pipped to the post by physicist Dennis Gabor. Check out the full investigation at https://biturl.top/yA3aYv

tourism is beginning. Back in the 1930s, when Norman Bel Geddes was dreaming up driverless cars, and Alan Turing accidentally invented AI, an iPhone would have been beyond anyone's wildest dreams. And right now there's probably one in your pocket (there is in mine). These aren't future technologies because they're here now.

I am writing this book using technologies that took decades to achieve. I'm typing on a Lenovo laptop with a 4GHz Intel Core i7 processor; in 1960, the PDP-1 ran with orders of magnitude less (about 187kHz) but managed to kickstart hacker culture, academic computing research and even gaming with *Spacewar!* There's plenty of room on my laptop's seemingly endless storage for my words, but why bother when I can access my files anywhere with Google Docs – a web app accessed via always-on, ridiculously fast broadband, that's automatically checked for spelling and grammar using an AI tool. A robot vacuums my floors, my family car can park itself without human intervention, and I can tell you exactly how many steps I took today (not enough). Living in the future is excellent.

And yet, where's my flying car – or my driverless one, for that matter? There's a swathe of future technologies (that I've been writing about for two decades as a technology journalist) that have long felt on the precipice of arriving yet remain just out of reach.

In the UK, the government proclaims annually that driverless cars will be on our roads this time next year* and driverless developers have promised we're just a few years from ditching steering wheels and pedals since Google's self-driving project first began in 2009. But what does that mean? First, companies were allowed to test them on public roads. And then, a year later, beneath almost identical headlines, driverless technologies inched forward to *advanced* trials. And so on.

* You can read my story for *WIRED* about this annual tradition here, but know that the UK government has continued to make this announcement each year since: https://biturl.top/raaqU3

In 2023, the year I write this, hands-free self-driving systems on public roads are finally approved for sale! But it's only level 2 (we'll get to what that means later) and a single car: Ford's Mustang Mach-E. That 'BlueCruise' automation system means the car can keep itself in lanes, avoid obstacles, and accelerate or brake when needed – and monitor the driver to ensure they keep their eyes on the road at all times. In other words, driverless cars are here … and they require a driver.

Progress is disappointingly slow because building things is hard, especially things we've never built before. The future technologies explored in this book all have a valid scientific grounding – they work in the lab and make sense according to our understanding of physics – but when engineers try to bring them into the real world, it's one heck of a lot harder. Virtual reality is remarkable, offering up whole digital worlds to explore – unless it makes you vomit. Hyperloops work in simulated models, but building one requires a route-length concrete tube propped up on towers in the sky. Driverless cars must contend with dogs darting into the road, activists plonking road safety cones on their hoods, weather that blinds their sensors and so on. One day, AI might evolve into that superintelligence we keep being warned about and be capable of handling everything we throw at it, including driving and robots. But in the meantime, until (and if) those breakthroughs happen, AI's utility remains limited.

Sometimes, the challenges are more human than technical. The first flying car was approved by regulators in the US in 1956 (yes, really) and yet, we keep whining that cars are stuck on roads. As it turns out, flying cars make more sense in sci-fi than in the real world. The same goes for cyborgs: most of us only want to replace what we've lost with robotic prosthetics and implants, rather than creating additional abilities. Smart cities sound exciting and efficient but fail to deliver anything but surveillance.

So why does it always feel like these technologies are just over the horizon? We humans are optimistic and enthusiastic, but these technologies are also overhyped. Too often, we're fed marketing rubbish rather than reality. Think of robots, and you're likely picturing Boston Dynamics' mechanical menagerie. The online videos suggest they can outrun us, do backflips and even dance with rhythm, so why can't they unload your dishwasher or scrub your toilet? That's because the robots were meticulously programmed to perform step-by-step instructions just for those theatrical videos. There is automation and intelligence in their movements, as taking a single step needs smarts – just ask the parent of a toddler. But these marketing videos reveal a far-off potential future for these machines, not what they can do now without human direction.

These futuristic mirages feel possible now because our technology seems so advanced – and it is. But we've felt this way before. Radio was deemed so miraculous it was considered as a means to control cars. The advent of virtual reality happened before we nailed computer graphics. The first radio frequency identification (RFID) sensor installed in a human wasn't a chip but a breakable glass vial. We never stop being impressed by modern technology, and can never imagine how much better it could be – or needs to be to bring these ideas to life.

Many of our modern technologies, and a few futuristic ones explored in this book, hinge on artificial intelligence. While there's no question that AI is more powerful than ever, we don't yet have artificial *general* intelligence: the ability to make a machine think. And perhaps we never will, or maybe it will happen several decades from now – after all, it took that long for AI to get to this point.

Technological development, particularly that associated with computing, feels like it moves at a freakishly fast pace, and it does. But take a step back, and it's clear much groundwork was laid to get here. Every Silicon Valley product launch was preceded by years of careful engineering, decades of science and just shy of a century's worth of advances in computing. Breakthroughs seem to happen almost overnight, but that is never the case.

Indeed, the rise of computing enables much, though not all, of the technologies highlighted in this book, whether it started that way or not. So it matters when advancements in computing happened, as well as where. Certain inventions had to arrive before others could follow: faster and cheaper computers enabled everything, while neural networks took off thanks to Graphics Processing Units (GPUs), wearable computers required lithium batteries, driverless cars were lost without satnav, smart cities needed connected sensors and so on.

★ ★ ★

Computing began in the 1820s with Charles Babbage and Ada Lovelace, who managed to design the world's first computer and write programs for it, respectively, though the steam-powered Difference Engine was never fully completed in Babbage's lifetime.* So much for the idea that the only way to predict the future is to build it. Alan Turing did the same in the following century, inventing core aspects of modern computing as well as AI before either could exist in the physical world. Imagining something new is a feat, especially when it's impossible to build and involves weird maths.

During the 1940s, mechanical computers arrived. They were massive hulking machines that ran computations using vacuum tubes and switches, including the 30-tonne (66,14lb)

---

* Daniel Crevier writes in his history of the subject, *AI: The Tumultuous History of the Search for Artificial Intelligence*, that Ada Lovelace, daughter of the famous poet Lord Byron, is perhaps unfairly credited with inventing computer programming as her ideas were never tested because Babbage's machines remained unbuilt due to technological limitations. But given that we credit others, including Alan Turing, for having ideas on paper before they could be brought into the real world, I don't see why she should be punished for Babbage's failures. Undaunted by their inability to make this computer work, Babbage and Lovelace instead invented games machines and a system to choose winners in horse racing; it didn't work, and Lovelace lost so badly she had to pawn her jewellery.

Electronic Numerical Integrator and Computer (ENIAC), built in the US to run calculations for firing artillery and later used for developing nuclear bombs. As the machine lacked memory, each new set of calculations had to be manually programmed, led by a team of women who designed algorithms and programs based on blueprints, as the high-level classification meant they weren't allowed in the same room as ENIAC.[*] At the same time, over in Britain, the Bombe was decrypting German messages during the Second World War, while afterwards, the Ferranti Mark 1 was the world's first commercially available computer, and tea shop company Lyons had the first business computer to efficiently dole out its cakes.

ENIAC had a core problem: the machine had to be rewired to change the program. Hungarian-born Princeton academic John von Neumann had a solution: store the instructions the same way you do the data. A machine that uses the von Neumann architecture – which is now all computers – fetches the instruction, then the data to be used and then performs the task, repeating as required.

In 1947, researchers at Bell Labs, based just outside New Jersey and led by William B. Shockley, developed a smarter way to do this using transistors. However, this Nobel Prize-winning genius was a pain to work for, so eight of his employees[†] left to form their own company, founding Fairchild Semiconductor and giving the local area the moniker Silicon Valley. Fred Terman – head of the engineering school at the nearby Stanford University – encouraged students to start their own companies, beginning a virtuous cycle of tech invention and reinvestment that was unmatched anywhere else in the world.

---

[*] Specifically, the team of women working on ENIAC was Kathleen Antonelli, Jean Bartik, Frances Holberton, Marlyn Meltzer, Frances Spence and Ruth Teitelbaum.

[†] One of the eight was Gordon Moore, the co-founder of Intel and creator of Moore's law.

By the mid-1950s, computers could use tape instead of punch cards for data storage and input. And by the 1960s, they started to do more while costing less – and taking up much less room. The arrival of the PDP-8, PDP-10 and 'microcomputers' like them spread computing technology through universities and businesses, with programming languages beginning to emerge, such as Fortran and BASIC (Beginners' All-purpose Symbolic Instruction Code). Artificial intelligence was coined as a term, programmable robotic arms were put to work building cars and the first message was sent over the Advanced Research Projects Agency Network (ARPANET), between the University of California, Los Angeles and Stanford Research Institute (SRI)*.

In the 1970s and early 80s, computers got personal, with Xerox PARC's Alto offering a graphical user interface, followed by personal computers (PCs) from Commodore, Tandy and Apple. A decade later, computers went portable with the Osborne 1 – though it weighed 11kg (24½lb) so was hardly a laptop. More powerful chips meant smaller computers, but also bigger ones, and the Cray 1 supercomputer was installed at Los Alamos National Laboratory in 1976.

In 1984, computers started to look more like their modern descendants. Apple introduced the Macintosh, with a mouse and a display – it's hard to imagine computers without these now – and a year later, Microsoft unveiled the first version of Windows. In 1989, Tim Berners-Lee outlined a plan for a system that became the web. And in 2007, the iPhone arrived. Now, chips are tiny, everything is connected and computing is ubiquitous.

That's the short version of computer history. But who paid for it all? Those first computers were war machines, and

---

* SRI International, as it's now known, was founded in 1946 as the Stanford Research Institute, a spinout of the famed university established to conduct R&D and consulting, and build partnerships. It's the creator of everything from the iPhone's Siri to Tide detergent.

there's no question that much of what we now see as computing and technology was initially funded by the military, including Fairchild Semiconductor. The US military realised academics could help the fight, wherever it was and whomever it was with, and so set up the Office of Scientific Research and Development, consultancy RAND and the Defense Advanced Research Projects Agency (DARPA), originally known as ARPA. The latter famously funded ARPANET, the precursor to the internet, and some of the projects we will explore shortly, including the Shakey robot, the driverless car challenge and Boston Dynamics' robots.

In the 1960s and 70s, the pushback against the Vietnam War sparked a backlash against military funds in academia, though US government rule changes that tightened budgets at DARPA and other agencies likely led to larger cuts than any protest. A successor was waiting in the wings: venture capital (VC) firms had started to evolve into the funding powerhouses they are today, beginning in 1946 with the American Research and Development Corporation (ARDC), the first to bring together cash and expertise.

The idea was to shift investment away from *who* you know – hopefully a Rockefeller or Vanderbilt – to the value of your idea. ARDC was an early investor in Digital Equipment Corporation (DEC), the maker of many computers that powered the research we're about to explore. By the 1960s, VC had reached the Bay Area thanks to Arthur Rock,* who advised the "traitorous eight" (as the Bell Labs breakaway group became known) to set up Fairchild Semiconductor rather than merely find new jobs, and founded the area's first VC company, turning $3 million into $100 million. Those successes were shared by the VCs, company founders and early employees, who would reinvest in themselves or other startups, keeping the money flowing – and growing. Indeed,

* Arthur Rock knew how to pick them, investing in Intel and Apple in the early days.

92 companies can directly trace their origins to Fairchild itself, including Apple and Advanced Micro Devices (AMD).

Of course, that's just Silicon Valley. Contrary to what many think, plenty happens outside of the Bay Area. A few names that pop up repeatedly throughout this book – Carnegie Mellon University, Massachusetts Institute of Technology and, further afield, the universities of Munich, Germany – have become research and innovation centres. These places act like magnets, attracting talent, ideas, money and media coverage. But we've also got projects running through the marshes outside Cambridge, England, driverless cars crisscrossing Europe and a whole generation of humanoid robotics in Japan. It doesn't matter who or where you are, you can try to build the future. And whether you're running a Stanford AI lab or blowing your retirement fund in your garage, you can fail all the same.

And it's these stories of engineers toiling against the gravity of failure that are so compelling. I've spent two decades of my career listening to marketing guff of PR departments pushing the next big thing, only to see it languish as an overhyped non-starter. How does this happen? Why is it so hard to take scientific breakthroughs and use them to engineer machines that can see, fly or think? Whether it's students on summer projects at prominent universities, mythical labs that have churned out dozens of successes but many more failures, or one-person shows solving problems one by one, these not-quite-yet technologies are built (or not) by people with dreams of what the world could be.

Some of them are wrong, often hilariously so. Still, it's easier to be a cynic and (accurately) point out the flaws than it is to engineer solutions, so they have my respect even amid their failures – well, except for the flying car test pilot who crashed because he was looking at the wrong fuel gauge; I'm just laughing at him.

CHAPTER ONE

Driverless Cars

Reinhold Behringer, a PhD student, is at the wheel of a Mercedes-Benz S Class saloon. Behind him sit dignitaries he's picked up from Charles de Gaulle airport. He guides the Benz on to the A1 *autoroute* towards Paris, flips a switch on the car's dashboard and stops driving. The onboard computer is now in control, with 11 miniaturised cameras acting as its eyes, letting this driverless car pass others, change lanes and maintain a safe distance from the vehicle ahead – all at speeds surpassing 130km/h (80mph) on public roads. Indeed, the other drivers on this motorway have no idea the car speeding by them is driven by a computer.

This wasn't a 2020s feat by a Silicon Valley tech giant. It was 1994 and the driverless technology was developed by academics at a Munich university. Forget Google and Tesla and their famous CEOs. The first driverless car hit public roads thanks to a German scientist named Ernst Dickmanns.

We've been dreaming about handing cars over to machines since we first took the wheel. There were mockups of systems that worked in the 1930s, vehicle-guiding motorways in the 50s and 60s, Japanese computer-vision driving in the 70s, Dickmanns's success in the 90s, and DARPA's challenges a decade later. While we feel at the precipice of driverless cars now, we've felt that way for decades – coming up on a century.

However, exactly what we mean by 'driverless' has shifted throughout the decades. Until recently, such automation was intended for motorway use only, taking the wheel at higher speeds but in more controlled road situations. Now, driverless cars are also being designed to work on side roads, meaning they must contend with pedestrians, pets and other confusing conditions. To differentiate, the American Society of Automotive Engineers (SAE) has developed an autonomy spectrum that begins at level 0 and goes up to level 5. Have an ancient banger of a car? Tell people you have a level 0 driverless car. You're not lying; it simply means there's no driving automation at all, although such a car may nevertheless include handy features such as automatic emergency braking or lane-departure warnings.

Level 1 includes driver assistance, such as adaptive cruise control or automatic steering for staying in lane; but only one or the other. Combine the two, so the driver gets steering and acceleration support, and you're at level 2. At these levels, the driver still needs to be actively managing the car all of the time, but software can offer support. Level 3 is a considerable leap ahead: at this point, the car gains the ability to detect the exterior environment – it has cameras and sensors – and make its own decisions. At level 3, the car can ask the human driver to take back the wheel; think of Tesla's Autopilot mode and Mercedes-Benz's 'Drive Pilot'.

Level 4 is where a car can manage an entire journey, but only in specific situations. For example, driverless cars in San Francisco can only operate in limited areas; you can't take them to many parts of that city, let alone to Canada. And even if the cars don't have a human safety driver in the front seat, engineers will be watching and assisting remotely if the system is confused, perhaps by sudden extreme weather, a waving traffic cop or a blocked junction.

Finally, level 5 is the possibly mythical car that can go anywhere, anytime, without human support, although whether it will ever be built remains to be seen.

★ ★ ★

Driverless cars were born at the 1939 World's Fair in New York. Fittingly for an expo held on the heels of the Great Depression, the theme looked to the future: 'Building the World of Tomorrow'. The 4.9 square kilometers (1.9 square miles) site in Flushing, Queens was formerly an ash dump – quite the real-life metaphor.

Opening just a few months before the Second World War eviscerated the world's optimism, the World's Fair featured pavilions highlighting the accomplishments of countries far and wide – including the Soviet Union, Poland and Czechoslovakia, though Germany opted not to attend. Think Disney World's Epcot Centre without the stands selling Mickey Mouse paraphernalia,* with displays of corporate-sponsored futurism alongside patriotic exhibits, all designed to entertain families on a wholesome day out. The fair was popular, opening on 30 April 1939 to more than 200,000 waiting visitors and drawing more than 44 million visitors over the two years it was open (aside from a temporary closure when the war kicked off).

One of the busiest exhibits was Futurama,† a delightfully named section of the Transportation Zone, set across the Grand Central Parkway to the west of the rest of the World's Fair (the spot is now home to Queens Zoo). Sponsored by General Motors, Futurama drew thousands of visitors every day, who waited in snaking queues to clamber into comfortable seats in a theme park-style ride that soared for 16 minutes above a 3,200-square metre (35,000-square foot) scale model of an American city, complete with half a million buildings and 50,000 cars, offering a glimpse of the 'motorways of the world of tomorrow'. The 'tomorrow' imagined by the exhibition was the 1960s, according to famed industrial and theatre designer Norman Bel Geddes. He believed everything on display would be accomplished within the next two decades (though his foresight didn't extend to predicting the war).

The exhibit illustrated Bel Geddes' argument that cars weren't living up to their full potential. In *Magic Motorways*, his 1940 book about Futurama, Bel Geddes noted that cars

---

* There was a Walt Disney-sponsored theatre, though. You can't escape the mouse. https://biturl.top/FfYfea

† Yes, the popular cartoon show *Futurama* is named after the exhibit.

could reliably drive up to 135km/hr (85mph), but congestion meant the average speed of motor traffic in the US was more like 32km/h (20mph). He also believed drivers were inherently unsafe, with accidents a common occurrence. There were 38,142 annual road deaths in the US in 1941 (compared to 2019, when 38,800 lost their lives on US roads, despite the population having increased by 150 per cent), so you can see his point. He was certain 'we are due to open our eyes any day now and demand a comprehensive basic solution' rather than accepting 'hit-or miss, spasmodic road "improvements" and catchy safety slogans'.

Bel Geddes' expo-based thesis argued that building a network of motorways was the only solution, replacing roads that evolved from the horse and buggy era with those specifically designed to suit cars. Instead of winding lanes, arrow-straight highways would allow cars to reach their full potential and speed, but safely. Those roads would need to be built from scratch, replacing existing infrastructure. In *Magic Motorways*, the ideal he points to enthusiastically is the Hudson Tunnel in New York, which segregates cars from other traffic, has no crossroads, keeps vehicles in separate lanes and encourages driving at a constant speed. These days, few people would admire it as a traffic-curing marvel. Still, those driving conditions were his aim – and that vision may sound familiar to followers of Tesla CEO Elon Musk, with his Boring Company wanting to drill tunnels beneath cities for congestion-free in-city driving.

Like Musk, Bel Geddes had another idea, one which dovetailed neatly with bespoke motorways: driverless cars. He predicted that 'cars that are automatically controlled, which can be driven safely even with the driver's hands off the wheel' would arrive by the 1960s.

It may seem astounding that driverless car technology was a subject of discussion in the late 1930s, but cars were increasingly adding automation. We've become accustomed to remote starters and locking, machine wipers, and tail lights that flash when brakes are applied, but all those technologies had to be methodically developed to make cars safer. But while cars were steadily improving, argued Bel Geddes, drivers were the same as ever. '[The driver] is still, day in, day out, on three million miles of road, the same,' he wrote. 'His eyesight is no better, he reacts no faster, he doesn't think any better, he gets drunk just as easily, he is just as absent-minded.'

Why, Bel Geddes asked, would we depend on humans to drive well when we can automate our way out of the problem? For example, Bel Geddes said that rather than encourage people to dim their bright headlights when an oncoming car approaches – a warning they may forget or fail to heed – the cars should sense each other and automatically adjust their lights.

Computers didn't exist in 1939, and 'artificial intelligence' was a term yet to be coined, let alone a developed technology. But other technologies, cutting-edge for the time, were believed sufficient to replace humans at the wheel. Bel Geddes' idea was to embed an electrical conductor into the motorway surface to guide a car via coils on the undercarriage, keeping it centred in the lane and an appropriate distance from the one ahead. There would have been no need to retrofit such technology: the interstate highway system across the US wasn't yet built, and the cables could have been embedded into roads from the start.*

Here's how he imagined the system would work for a truck. A human driver is at the wheel on minor roads, only giving up the controls when entering a motorway feeder lane, when the automation takes over. The truck automatically speeds up for the motorway, other vehicles alter their speeds to create a gap in the flow of traffic and the truck enters to find its spot in the lane. All vehicles travel at the same speed, maintaining the perfect distance between them, and the driver can sleep

* Bel Geddes wasn't tied to a particular technology, also suggesting radio, as it was already used in aeroplanes for guidance. Stations along the highway could beam a radio broadcast at cars to control traffic, he suggested.

or stare out of the window – he's not needed. When it's time to depart the automated motorway, the driver taps a button on the instrument panel and the vehicle manoeuvres into a slower lane ahead of its exit. 'You do not need to use your horn (with which your car is equipped for local roads) on this motorway because you never have to pass a car,' he wrote.

Of course, all Bel Geddes did was imagine driverless cars, as sci-fi films have done many times since. Bel Geddes didn't build a driverless car – though he very much thought it was not only possible but also practical.

And he was right. Though interrupted by the war, the daydreaming ideas on show at Futurama were put into practical development, and in 1957 the first driverless car was tested just outside Lincoln, Nebraska.

After the war ended, Bel Geddes' idea of removing human mistakes from motorways via automotive automation was revived by General Motors, alongside staff from the labs at the Radio Corporation of America (RCA) and the Nebraska Department of Roads. The aim initially was to build a system for safer driving in reduced visibility, with automated controls seen as a future goal, as that required additional infrastructure in the car and on the road. RCA researchers, including famed engineer Vladimir Zworykin, built a miniature version of a cable-guided car – the same design tourists and families marvelled at in 1939 – on the lab floor in 1953. If you think that sounds like a Scalextric slot car set, you're absolutely right. What we would call a toy was dubbed a 'small-scale system' and grabbed the attention of Leland M. Hancock, a traffic engineer in Nebraska.

The RCA's magazine, *Electronic Age*, described the concept as similar to the Futurama idea: drivers manage side roads, but pass over control to automation for motorways with the press of a button. The car manages the speed and keeps within its lane, while the driver 'may prefer to read or carry on a conversation ... it makes no difference for the next several hundred miles as far as the driving is concerned'.

The actual trial didn't run for hundreds of miles. In 1957, the necessary hardware was embedded in just one 120-metre (130-yard) section of road, just outside of Lincoln, where a new intersection was under construction. Circuits embedded into the asphalt kept a Chevrolet equipped with sensors centred in the lane. At the same time, speed was calculated using lights along the edge of the road while radio receivers inside the car controlled the steering, braking and acceleration.

There were three parts to the system. First, detectors were installed along the road to count and track vehicles; these were loops of wire buried in the pavement and connected to a circuit at the edge of the road. When a car drove over the buried wire, it was picked up by the roadside circuit as a change in current, and used to control a warning light along the side of the road.

The second aspect was a radio warning system. This was a basic transistor switch with an antenna embedded in each of the detector loops in the road. When a car passed, the transistor switch closed and the antenna sent a signal at a different frequency to the first system. This radio signal kept other cars from getting too close by alerting them to apply the brakes. The third part of the system was a cable laid in the centre of the lane, emitting a signal at a third frequency. Cars with receiving antennae mounted on the front and back could follow the signal. As the car moved too much to one side, it would self-correct and shift to the other.

By all accounts, it worked. *Electronic Age* reported the car could automatically warn a driver that they were following another car too closely or there was an obstacle in the road, while the driverless version could follow its traffic lane even with zero visibility. However, the lane control wasn't fully trialled in 1957, just shown on a dial in the car.

Two years later, the RCA team showed off fully automated cars on a quarter-mile test track in New Jersey, offering rides to reporters. It was the same system as before – cables in the road, sensors on the bumpers, radio announcements about exits – using a General Motors car with no steering wheel or brakes, just a simple joystick.

In a story that wouldn't be out of place in *WIRED* or *The Verge* today, the local *Princeton Press-Courier* ran an article headlined 'Reporter Rides Driverless Car', with the journalist, Doc Quigg, describing his ride in the car with 'no hands, no feet, and no brain operation'. The test car successfully stayed 25–30m (80–100ft) behind the other on the track, stopping when the lead vehicle halted. Then the demonstrators decided to show off a bit, leaving a car parked on the far side of the track as an obstacle for test cars. Quigg noted the demonstrator saying: 'Let's suppose we're cruising normally down the highway around a blind curve and we don't know a car is stopped in front of us and – oops!' The demonstrator had to slam on the brakes himself and take back manual control of the car; he'd forgotten to flip the necessary switch to enable the automated mode. 'Nobody screamed,' noted Quigg. 'But the age of automation can have its moments.'

And, as now, journalists were eager to point to flaws in the system. Quigg noted that the road-embedded circuits only registered the presence of any metallic vehicle, as well as its speed and direction. 'RCA engineers admitted that a herd of cattle would not be detected, unless they were dressed in armour – but then, the limited access highway of the future will be closed to cows,' he quipped. So we could have had driverless cars in the 1950s, but only on motorways clear of everything else, be it cows or other non-metallic objects.

The idea of cables in roads to control cars on motorways also spread to the UK, with the government's Road Research Laboratory (RRL)* developing its own version of magnetic controls, which also used embedded cables in the road and sensors in the car. It did so using a 1960 Citroën DS19, a car

---

* The RRL subsequently became the TRRL (Transport and Road Research Laboratory) and is now a private organisation known as TRL, a pattern of privatisation common in the UK.

so unique that when RRL offered its driverless model to the Science Museum in London for its collection, the museum replied that it would have accepted the car even without the automation research because of its advanced and unique hydroelectric suspension system.* That self-levelling suspension system wasn't necessary for the automation work, but the car's power-assisted steering system meant it could be easily modified for driverless modes, though later electric-assisted steering systems proved less complicated to automate.

Two sensors, one on each of the car's front corners, detected the magnetic field of the current running through the cable buried 70mm (2.75in) deep in the middle of the road. The difference between the measurements of the two car-mounted sensors revealed whether the car was deviating from the centre of its lane and by how much, and also calculated its speed, using this information to ease the car gently back into position for a smooth ride. The control system sat in the passenger seat, with plastic embossed labels noting which switches applied to the sensors and which to the differential system. In the middle of the dashboard was a red button, a label above it reading 'manual override'.

The steering actuation system turned the front wheels without turning the steering wheel, which was disabled in automated mode as it interfered with the driving. A clutch in the steering system let the driver flip between automated and manual mode, effectively turning the steering wheel off and on, made easier by the Citroën DS19's power-assisted steering.†

The car worked. It could top speeds of 130km/h (80mph) on the straight without shifting more than an inch from the centre of the lane, though it slipped as much as five inches

---

* The car remains in the collection at the Science Museum in London, UK, though it currently isn't on display, it has been loaned out for car shows.

† This is why researchers used a car with a joystick in one version during the US trials.

through corners at such speeds. Slowing improved accuracy and would be advisable on a tight turn anyway. This issue of keeping the car in the correct lane position through steering was known as the 'lateral control' problem.

There were other problems too. The car had to slow down in snow and rain, topping out at 90km/h (55mph), though the researchers noted that human drivers couldn't do any better. The asphalt test track also had two sections of steel-reinforced concrete, and the metal interfered with the cable signal.

Additionally, while the cable kept the car centred in the lane, it's no good to have a car perfectly inside its lane if it rams the car in front or holds up vehicles behind. This was the 'longitudinal control' problem: how fast should it go, and how would it know other cars' locations?

There were two solutions. The first was 'synchronous slot', also known as 'the travelling bucket', in which the track has a moving electronic reference point for the cars to ride, like a surfboard on a wave or suitcase on a conveyor belt. The second was 'relative headway', or 'follow-the-leader', where each car's position is in relation to the car ahead, requiring constant measurement. Travelling buckets are good for junctions, as cars can be clumped together to avoid stopping, but if one breaks down, others might be held up, meaning an additional backup control system would be required. So RRL opted to play follow-the-leader, with lasers, radar and ultrasonics all candidate technologies for assessing the distance to other cars. One idea to use radar was deemed likely to have worked but would have required a radar system on all vehicles on the road – even those that weren't driverless.

Either way, the researchers noted that electric motors would make the job easier, as they'd allow roadside transmitters to communicate with the cars, telling them to speed up or stop if necessary – this was one reason the researchers switched to an electric Mini and Ford Cortina; the other was complaints about the use of an expensive 'foreign' car. To achieve the desired speed benefits, the cars must keep close to each other, but that raised other issues. 'It also seems essential for public acceptance of automatic cars that any vehicle should always be able to brake to an emergency stop without being rammed from behind,' noted Slade Penoyre, principal scientific officer at RRL, in a paper delivered to attendees of the Journée D'Étude sur le Pilotage Automatique conference held in Paris in 1972. Yes, indeed.

The aim was that all cars would have 'auto-drivers' similar to aircraft autopilot systems. 'The vehicles would be driven in the normal way on minor roads but would be switched to the automatic mode on major roads and motorways,' according to an RRL document. 'This mode could include automatic route finding as well as speed and steering control.'

Initially, it was forecast that such a system would cost £75 (£1,000 today) per car, though media reports later suggested the actual cost would be closer to £200. However, as the system never went into production, it's impossible to give an exact figure. But even £200 is a small slice of the total expense of the car, with the DS19 selling for £1,700 in 1969; that said, this doesn't include the cost to the taxpayer of embedding the cables in the roads.

Why go to all this effort? Fewer accidents and also less traffic, as the accuracy of machines over humans would mean more lanes could be crammed on to existing roads and cars could drive faster with less space between them. Such vehicles would also eventually be cheaper and more efficient than public transport. In the same paper delivered in Paris, Penoyre noted that the benefits would combine those of car and train transport, including 'door-to-door movement, no waiting for or changing vehicles, privacy, ... safety, high speeds, [and] no driving strain'.

The RRL researchers laid out a long-term road map for rollout that would see widespread use by the turn of the millennium. First, automated buses would operate on separate tracks, followed by limited private cars and delivery vehicles, also on segregated roads. Then, some motorways would be given dedicated lanes, with the necessary cable embedded in the tarmac to be used by sensor-equipped cars, while 'freight trains' of lorries led by a single driver would be allowed access at night. Slowly, more lanes, motorways and vehicles would be equipped.

But that date has come and gone, so why does the UK not have cables embedded in roads? RRL continued to work on the project for a number of years, trialling the system on a bus as well as the Mini and Cortina. However, the economic collapse in the 1970s meant funding disappeared, and RRL ended the project.

Indeed, high costs are the main reason we don't have automated motorways. In 1969, Robert Fenton and Karl Olson of Ohio State University published a paper on solutions to create driverless cars. Though they identified a whopping 1,296 different ways to make this happen, their proposals also relied on embedding electric circuits directly into motorways, with drivers taking back control on more minor side roads.

Fenton and Olson noted in their paper, 'The Electronic Highway', that the costs would range from $20,000 to $200,000 per mile, alongside the car, which would cost several hundred dollars to kit out. Adjusted for inflation, that upper figure is roughly $1.7 million per mile. The cost of building a multi-lane motorway in the US is already several million per mile, so, in short, the additional infrastructure made an already expensive undertaking even pricier for a technology that, so far, had only been proven in controlled conditions.

The total bill wasn't the only challenge; a bigger problem was who would stump up and pay? The cost was predominantly in roads rather than the cars themselves, meaning the money would have to come from the government's infrastructure budget. The way driverless cars are designed now, the costs are largely with the consumer: to buy a Tesla with an 'Autopilot' package – which is *not* fully driverless – you are paying an extra $6,000 on a base price of about $50,000, while 'Full Self-Driving Capability' – which, again, to be perfectly clear, is *not* fully driverless – is $12,000[*] plus a monthly fee. Aside from some regulatory considerations and clean-up costs when it all goes wrong, individuals pay rather than governments.

It's a shame the experiment failed, as the system employed technologies beyond driverless that could have changed our world, particularly around electric vehicles and recharging them. In 'The Electronic Highway', Fenton and Olson suggested that roads with embedded cables could also be used to power electric cars while they travel. And we're still working on this: tech company Qualcomm has outlined plans for wireless road charging, while Stockholm has installed a charging rail in the road to which electric cars connect for a brief recharge while still travelling. Imagine if drivers could also take their hands off the wheel for a quick break to recharge themselves.

Of course, driverless development remains expensive, with estimates suggesting as much as $80 billion[†] has been spent in recent years creating self-driving technology. Though the US government does shell out in the form of grants and education, and research at organisations such as DARPA, the vast majority of that spending isn't fronted by the government itself but by private investors looking for Silicon Valley's next big thing – and unlike governments, they aren't answerable to taxpayers.

So it's no surprise that the next round of attempts to enable drivers to take their hands off the wheel ditched the roadworks in favour of computers, which had quickly become small enough to fit in a car and powerful enough to drive one – if only just.

★ ★ ★

---

[*] As of September 2023.

[†] In 2017, the Brookings Institution attempted to tally up the cost to date, estimating $80 billion had been spent globally on both development and acquisition. More has clearly been spent since. https://biturl.top/JF3qEf

Here, we turn to Ernst Dickmanns, a German aerospace engineer considered to be the father of driverless cars. His mid-1990s system will sound like the definition of success to modern driverless developers: his Mercedes van 'saw' using cameras and sensors, with artificial intelligence churning through the data to make decisions. And his team took it on public roads at speed, covering thousands of kilometres.

Dickmanns was a professor at Bundeswehr University Munich, teaching computer vision and machine guidance courses after a career in aerospace research. It may seem an odd route, to go from space vehicles to driverless cars, but a robot sent to another planet to explore must be able to look around and navigate independently. If you can make it work on the moon, you can do it in Munich.

In the late 1970s, Dickmanns started work on *Versuchsfahrzeug für autonome Mobilität und Rechnersehen* (VaMoRs) or, roughly translated to English, 'Test vehicle for autonomous mobility and computer vision' – and it did just what it said on the tin. The first system used a Mercedes-Benz Vario van, a 5-tonne (11,023lb) beast of a vehicle, but the large size was necessary to provide space for the stacks of computing hardware needed to analyse images in real time. VaMoRs pulled data from its surroundings via multiple cameras, to be processed via those onboard computers.

It took eight years of work, but in 1986 the VaMoRs van was ready for a public test around the university campus. Videos of the van in action show the wheel turning on its own, the accelerator and brake depressed as if by ghosts. The next year it was sent down the soon-to-be-opened autobahn at a top speed of 96km/h (59mph) – as fast as the van itself could travel. Picture a vast, empty road with a Mercedes van hurtling down it, no driver at the wheel. Even in low-resolution black-and-white video,* it is thrilling.

* You can watch the video here: https://biturl.top/VBNZJr

VaMoRs was the first stage. The carmaker then known as Daimler-Benz (now Mercedes-Benz Group) began supplying vehicles. Dickmanns' research focused on computer vision; he managed these feats of driverless prowess using only cameras and a bit of modelling, believing that to be entirely sufficient. Daimler-Benz had its own version of the automation technology with added lidar and radar for additional safety, as it sought to eventually commercialise the idea rather than just to write a research paper about computer vision. Because of the extra sensors, the team needed an even bigger van to fit all the computing required.

The VaMoRs tests won Dickmanns and his team funding from a key driverless project. In 1986, European countries joined up their research into road safety via Eureka, a research collaboration group, working with car makers on a project dubbed Prometheus* – and it came with a massive €749 million budget.

Prometheus investigated various ideas, including conductive cables in roads to guide cars – 50 years after Bel Geddes' exhibition, the idea still had merit. But once they saw Dickmanns' 5-tonne (11,023lb) van hurtling down the autobahn, Eureka ditched the cables and threw the funding at the German engineer. Dickmanns set to work optimising the VaMoRs system, adding cameras and boosting computing, letting the system not only navigate and stay centred at the correct speed, but also to spot obstacles. That also allowed vehicles to recognise each other and drive as a convoy, with one following the other at a set distance, and to change lanes.

Thinking beyond vans, Daimler-Benz supplied passenger cars, a pair of Mercedes S 500s – 'a really nice 12-cylinder thing with leather seats', recalls Behringer. The seats didn't last long: the team ripped out the front passenger side to house

---

* Arguably, it should really be PROMETHEUS because it's an acronym for PROgraMme for a European Traffic of Highest Efficiency and Unprecedented Safety. But come on, that's ridiculous.

computers, more of which were also crammed into the rear seat and the boot. There was still room for a test driver and a co-pilot, who would sit in the rear, watching monitors.

When Behringer joined the team, the technology was already up and running, though there was plenty of work to go around with incremental improvements, shifts to different programming languages, hardware types and applying it to the new cars. So far, they had driven on an unopened stretch of the autobahn and on test tracks; next they wanted to take the system out in traffic on a public road. Navigation was an issue as GPS wasn't available. That meant creating a fully autonomous car – one to which you could simply say 'take me home' – wasn't likely to happen anytime soon. 'Instead, our short-term goal was to go onto the motorway, and once there, switch the system on and it would bring you safely to the exit where you wanted to get out,' Behringer tells me.

The trick was to do that at full speed. As Behringer recalls, plenty of autonomous systems were under development in the US at the time that used an AI technique called neural networks to move a vehicle or robot, but they were merely inching forward. 'Dickmanns was very against neural networks,' Behringer says, and though as a student he was disappointed at not getting to work on the intriguing new technology, he now agrees. 'He was more rule-based – it's physics.' He believed you don't need a neural network with an expansive model and years of training to tell a car to stay in the middle of the road – it's not something a car should have to learn. Just programme it with the rules of the road and an understanding of physics, and a car should be able to stay centred and avoid other vehicles. 'When there's a left marking or a right marking, it's very clear to drive in the middle,' Behringer says. 'If the car ahead is too close, you need to slow down, and the laws of physics help you achieve that in order to not collide.'

To be clear, Behringer isn't arguing against using neural networks, just against relying on them entirely. Indeed, the two systems can be combined, with the rules-and-physics side managing core driving aspects and a neural network extending its capabilities in unexpected situations.

The Dickmanns system was based on computer vision being interpreted and applied to that set of rules. Of course, lane markings can disappear or be interrupted, so the system needed to consider that too, predicting where they should be, with everything updated every 25 milliseconds. That was Behringer's job. Other students worked on object detection for cars and pedestrians, on translating commands from the computers to the car's controls, and on a way to hand back the wheel to the driver. And that's key: Dickmanns' system didn't pretend a driver could ever be fully removed from the system. This was a level 2 car, Behringer says, designed for motorway driving free of people and other potential dangers, not for side roads. 'Our driving mode was the motorway, because that's the low-hanging fruit – it's a relatively easy, constrained environment, because at that time, 30 years ago, computers were not that powerful,' Behringer says. 'You have to compromise, and we know in the future with Moore's law, they will get better.'

As processing power advanced, the team could use smaller computers, allowing for the switch from vans to those Mercedes sedans, known as VaMP and VITA-2.* They navigated via mapping and vision alone. Miniaturised TV cameras looked to the front, rear and side of the car. The images were analysed by 60 transputers, a new type of microprocessor designed for parallel computing that could be linked together into a network to build a miniature server farm. By the 1990s, those were dropped for newer microprocessors, as computing technology had leapt forward enough to be useful to driverless cars.

* VaMP was run by the university, and VITA-2 by Daimler-Benz. The latter had more cameras and sensors, but otherwise they were largely the same.

But the available processing power still wasn't quick enough to enable the cars' computers to churn through and understand the images pulled in by the mini-TV cameras.

The researchers had two solutions. First, they developed a system to reduce what they needed to look at, considering the edges of objects rather than their entirety. Second they only looked at aspects that changed or moved, and used modelling to make predictions about what would happen next to give the system a sense of how objects move in sequence. This reduced the time needed to analyse images coming in through the car-mounted cameras, and the same idea is still in use today.

Dickmanns called that system his '4D approach', and it worked. 'So the community was quite surprised when we demonstrated our first running on the autobahn in 1987 at speeds, and the maximum speed of our vehicle VaMoRs, at 96 kilometres per hour,' he said in an interview in 2010 for the IEEE History Centre, noting that vehicles in rival projects in the US and elsewhere were travelling at 5km/h (3.1mph), due to being limited by the rate of image processing. 'And this was only possible by developing these methods.'*

The team operated out of Dickmanns' institution, Bundeswehr University Munich, a military university located at a former airport. 'I had an office in the former [control] tower, with a view of the Alps from Switzerland to Salzburg,' Behringer says. They turned the disused runway into a test track by painting the necessary road markings. The system had to see the lane markings and understand how they curved by looking ahead, keeping the car to the centre. They hoped it would work as the system had been simulated first, though Dickmanns tells me via email, 'Of course, there was a feeling of happiness that our preparations had shown to be valid.'

* You can read the full interview with Dickmanns' here: https://biturl.top/Z7Vziq And the CHM's full collection of oral histories are here: https://biturl.top/aim2M3

Students did virtually all of the driving; Dickmanns tells me he never drove the car, though he did drive the van. Behringer would sit in the rear seat, monitoring the system. Test drives at the track made the student nervous, as his code was often tested: 'I make mistakes.' And the only safety equipment was basic – just seatbelts and airbags. But driving was the most nerve-wracking job. 'You don't even see the monitor, it's just looking at the road and looking at the steering wheel and trying to grab it as soon as there's jerky movement,' he says of testing at the track. 'We learned to observe some signs, for example when it seems to drift to one side, that is not normal.'

On the other hand, when the ride is perfect, that's also a problem. 'Then you get actually too much trust in the system,' Behringer says. It worked so well that they ignored the officially approved top speed of 130km/h (80mph) and edged over 180km/h (110mph) while on the public motorway.

In October 1994, after somehow convincing the French authorities that this was a wise and safe move, Dickmanns' cars were driven by researchers to Charles de Gaulle airport, where they picked up notable guests. When the cars reached the A1 *autoroute*, the researchers handed driving over to the onboard computers, which controlled steering, acceleration and the changing of lanes as necessary, up to speeds of 130km/h (80mph) amid public traffic – though the human drivers kept their hands on the wheel and had a red button to bash to regain control in case of emergency. There was nothing about the cars to indicate to other drivers on the motorway that they were driven by computers – if an S Class passed you in late October 1994 on the motorway from Charles de Gaulle to Paris, you may have been part of driverless history without ever knowing it.

Next, the system was tested via a road trip from Munich to Odense, Denmark, with the car successfully passing other cars and changing lanes at speeds surpassing 180km/h (110mph). There was, of course, a human test driver on board and the car didn't manage the entire trip on its own; in particular, it was tripped up by sections of the road that were under construction.

Still, the system handled 95 per cent of the driving, once travelling a 158km (98-mile) stretch without handing the wheel back to the researcher, and clocking up an average distance between interventions of 9km (5½ miles). The system managed the distance between cars, speeds and lane changes, though the latter were still manually triggered as the car lacked a rear camera and needed the test driver to shoulder-check.

On that long route across Europe, the driver would invariably get bored and want to show off the system, so passed other cars with his hands off the wheel. 'But we might have gotten a police report because no one knew it was automatically driving, so that was not a good thing to do,' laughs Behringer.

Given that the system worked then, why weren't driving licences obsolete by the turn of the century? Well, as good as Dickmanns system was, more work was needed. It struggled to see negative obstacles, such as a hole in the ground. That was partially overcome by adding stereo vision – two cameras looking at the same point, which gives a depth measurement – and lasers. But the primary challenge was computing power: processors couldn't churn through the data fast enough. At the time, Dickmanns predicted it would take two-to-three orders of magnitude more computing power than what was currently available. That was expected to take until at least 2010. The lack of progress, plus the high costs, scuppered the academic side of the project.

Behringer points out that a car that can manage *most* of the driving is insufficient. 'We had 95 per cent – that's nothing. You need 99.999 per cent – every decimal takes another five to ten years,' he says. 'It wasn't a consumer product, it was a concept demonstration. We were fully aware of that.' Mercedes kept at it for 10 more years, took it 'a few digits further' and the company's 'Drive Pilot' was the first technology to win approval for driverless level 3 – it can be driven in selected parts of Germany as well as Nevada and California, so long as you're willing to shell out €5,000 extra on your S Class.

Dickmanns says full driverless will happen – just give it time. 'I am convinced that this technology will be realised in this century,' he tells me via email. 'The development of safe software and of specially designed cars will take quite a bit of time; fast and early quick shots are not the way to go.' His work continued after the European project, teaming up with American researchers to apply his techniques to minor roads and even off-road driving. Dickmanns' approach continues to be developed at a German institute, Technik Autonomer Systeme (TAS), led by a former student.*

Others beyond Dickmanns were also trying to make driverless work. In 1977, engineers at the Tsukuba Mechanical Engineering Lab north of Tokyo sent their driverless car crawling towards a stop sign in the hope it would stop. It did, using dual cameras interpreted by onboard computers; however, processing speeds were so slow they not only limited the car's speed to 30km/h (18mph) but meant the researchers had to add extra road markings to assist the car. In 1991, US Congress funded a project to examine automated roadways using a 12km (7½-mile) stretch of California highway as a testbed. This saw cars controlled in a variety of ways, including cameras trained on reflective strips for lane centring, radar for keeping cars spaced correctly, and even magnets embedded in the road for guidance – all ideas previously trialled, but none at the same time on a major highway. And in 1995, two researchers from Carnegie Mellon University (CMU) drove from Pittsburgh to San Diego using a partially driverless car; they called the nine-day trip 'no hands across America'. The NavLab 5 system used RALPH (Rapidly Adapting Lateral Position Handler) to analyse video imagery to keep the Pontiac Trans Sport van on the road and in the right lane,

---

* Behringer believes that the wider public largely forgets Dickmanns' efforts, though those in the industry know his name, and the scientist was awarded *Bundesverdienstkreuz*, the German order of merit, for his work.

which it managed 98.2 per cent of the time, though the researchers had to manage accelerating and braking.

In 2003, DARPA issued a challenge which heralded the rise of AI-based driverless systems: create a driverless car, win our race, and we'll give you $1 million – surely the most expensive, difficult way to become a millionaire, not least because the route wound through the Mojave Desert near Barstow. Why? Because DARPA didn't want driverless cars to ease traffic but to use as military vehicles to help soldiers avoid roadside bombs in Iraq and elsewhere in the Middle East.* Forget a cable-controlled classic Citroën blasting around a British track, or a computer-seeing Benz dashing down the autobahn – the first across the finish line was a VW stop-starting along a dusty lane. And even that took a second try.

No team won the first year's staging of the DARPA Grand Challenge. In fact, none even completed the 228km (142-mile) course and half of the 15 cars taking part failed to get further than a mile. A motorbike called GhostRider, built by Anthony Levandowski, fell over at the start line because the team forgot to turn on the carefully designed stabilisation system. A team of high school students managed to cross the starting line, but their Acura SUV immediately drove into a barrier. One vehicle had the wrong GPS location programmed and tried to drive to Indiana. The entry from Dickmanns' student Behringer managed a respectable mile before a USB plug connecting the system's hard drive came loose, and the car ended its run.

---

* Initially, according to Alex Davies' *Driven*, the plan was to run the race from LA to Las Vegas, and that was what DARPA announced at the launch event. But they quickly realised they couldn't actually shut down major public roads to run the test, so shifted to the desert, though that too involved negotiations with local indigenous groups and an environmental impact plan. It's almost like they didn't really think the whole thing through.

The first vehicle out of the gate also went the furthest: Sandstorm, made by CMU's Red Team, and led by famed roboticist William 'Red' Whittaker, who had previously designed robots to clean up the mess at nuclear disasters including Chernobyl. Using lidar, radar and dual cameras, the Sandstorm Humvee managed to travel nearly 12km (7½ miles) of the total 228km (142 miles), though not easily: it spent two whole minutes pushing against a fence post – the military-style vehicle won that battle – before bonking into a rock and becoming stuck on an embankment, where it then spun its wheels until they lit up in flames and the team gave up.

Others found success in other ways. The spinning lidar on the top of a driverless pickup truck run by brothers Dave and Bruce Hall and team garnered attention – and their company Velodyne became a major provider of such systems to driverless developers in later years. They developed the code while commuting to work, one brother coding while the other drove. At first, they used stereo vision cameras, dual cameras that they intended would allow their system to calculate the difference between objects in the image to figure out what to avoid. When it didn't work, they ditched the cameras in favour of a lidar system.

Many in the media laughed, calling the competition overhyped. The *Register* said the 'RoboFlop' was a 'pretty humbling display' but admitted DARPA's money was well spent, bringing attention to robotics and letting the government pick from the best of the wreckage. Despite the apparent failure, DARPA was happy with the progress, particularly the number of teams that signed up, and doubled the prize money. However, the next year, the route was changed so the hardest terrain was no longer at the beginning but at the end.

The second time around, 195 teams signed up, so DARPA whittled down the numbers to 43 with a pre-challenge obstacle course at the California Speedway. It must have been one heck of a day for spectators: a 10-tonne (22,046lb) truck took eight tries to get past a set of traffic cones and a vehicle with the appearance of a retro metal rocket ship careened out of control into barriers. Levandowski returned for another try with his GhostRider motorbike, having worked on the project constantly in the intervening years, saying how he hadn't been seeing his friends and didn't have a girlfriend. 'It's worth it though; this could be a chance of a lifetime,' he said in a documentary about the race.

Among the successful teams to qualify for round two were Sandstorm and Highlander, both led by CMU's Whittaker, which between them had a pair of Humvees, a $3 million budget and a contingent of 100 students, making it the largest team. Each team's vehicle featured six laser scanners with specific roles, such as looking for the edge of the road or obstacles, as well as a long-range laser, radar, GPS and other sensors, with seven Pentium M processors churning through the ensuing data – a system that would have been remarkably powerful to the software engineers working with Dickmanns outside Munich in the 1980s and 90s. Whittaker's engineering students designed a bespoke gimbal to let a sensor spin in order to look sideways off the car, offering a longer-range view around corners. Whittaker decided to run Highlander fast and Sandstorm slower, giving CMU two chances to win – one through speed and another through care.

A new entrant for the second year was Stanford Racing, from the world-famous university located at the heart of Silicon Valley. The team was led by Sebastian Thrun, at the time the youngest person to head up Stanford's AI lab. 'I felt that field should be better than this,' he told me.

The best-performing vehicle from the first round, Whittaker's entry, wasn't looking at obstacles but at GPS points. And as Thrun notes, that vehicle drove over a big fence post. 'That should have shown up in the lasers, right? So it looked like the vehicle had no understanding of the fine details, and if you drive off GPS points, you can be a metre off easily – in this environment, it's the difference between life and death.'

The first time Stanford's entrant, a 2004 Volkswagen Toureg named Stanley, went out publicly, for the qualifying round, it was the only car not to hit an obstacle on the track.

Stanley had been donated by the carmaker, with actuators in place to control the brakes, steering and so on. Though the researchers still had to add sensors such as lasers, they bought off-the-shelf versions rather than designing their own. That allowed them to focus on software, in particular the machine learning that would let the car analyse the world as it went, with software churning through the sensor data and overlaying that data onto video camera footage. The laser data identified the smooth road directly in front of the car and extrapolated that route forward, updating the directions as the car advanced and collected more information. This adaptive system allowed for flexibility – if a fence post was spotted in the road, the car could be told to drive around it. 'We believed from day one that it was a software race, not a hardware race,' Thrun says. 'The car doesn't fail because it's not rugged enough, [but] because it's not smart enough.'* What surprised Thrun was how necessary machine learning was to cracking automated driving.

Three days after the qualifying round, the second Grand Challenge kicked off. The 212km (132-mile)-long course across the Mojave Desert had a 10-hour time limit. The route

---

* In a *NOVA* documentary (Season 33, Episode 9) on the DARPA challenge, Thrun is shown testing Stanley, with coder Mike Montemerlo, along the route of the first race. As they pass the location where CMU's Sandstorm got stuck, Thrun notes its history and says they're glad to get beyond that point – and almost as soon as he says it a software bug crashes their system and their journey ends exactly where their former colleague and now rival's car got stuck the year before. These vehicles may not be able to drive very far, but they have a sense of humour. You can watch it here: https://youtu.be/vCRrXQRvC_I

was handed to teams at 4am on the day, giving them enough time to upload the relevant route data to their cars before the race. DARPA vehicles followed each entrant, each equipped with a kill switch if necessary. Stanley needed to know the general route but would figure out exactly how to go about driving when it got there. The CMU Red Team used its massive army of students to quickly break down the route and tell its vehicles as much about how to drive as possible, including what speed to go.

That year, all the vehicles managed to leave the starting zone, and most surpassed the previous year's furthest record of 11km (7 miles), though not all finished. The Hall brothers' entry was knocked out after 42km (26 miles), when the rooftop laser was knocked loose. As Whittaker's Highlander came up a hill, its engine suddenly cut out, leaving the vehicle sliding back down. It recovered and kept moving, but slowly on vertical climbs; the gimbal had failed, further exacerbating the challenge.*

The fault gave Thrun's car the chance to catch up and, five hours in, Stanley passed Whittaker's Highlander. Highlander's sibling Sandstorm, tasked with taking the race slowly and carefully, was unable to catch up. After almost seven hours driving at an average speed of 30km/h (19mph), Thrun's team walked away with a win for Stanford, helped in part by Whittaker's woes. Nevertheless, Highlander and Sandstorm followed shortly after, nabbing podium finishes, followed by the Grey Team, fresh from post-Hurricane Katrina New Orleans, and TerraMax. With no other team finishing, it was clear the battle was between Whittaker's hardware-focused vehicles, with engineering students making their own gimbal,

* A decade later, the CMU team finally figured out what went wrong: a filter in the engine had a flaw, perhaps caused by the vehicle rolling a few weeks before the race. It meant that there was nothing wrong with the driverless system itself. You can read more at IEEE Spectrum here: https://biturl.top/jEvUBz

and his former colleague Thrun's 'cobble it together from-off-the-shelf bits but with brains' idea. Both worked well.

By 2007, the competition had moved out of the desert and on to city streets – well, the disused George Air Force Base in California, anyway. Instead of dusty deserts, the cars were tasked with managing intersections, parking and merging, all while manoeuvring past each other as well as 30 vehicles driven by humans. The addition of traffic was the main difference, says Thrun, noting that in the previous years in the desert, if cars came close to each other, one would be asked to stop to let the rival pass safely. 'In reality, the traffic was moving very slow,' Thrun says. 'It wasn't like Manhattan or something like that.'

This time round, Whittaker finally beat the competition, giving CMU a much-deserved win with an average speed of 22.5km/h (14mph), pipping Stanford Racing's runner-up, Junior, which completed the course at 22km/h (13.7mph). But all of the vehicles managed to manoeuvre the setting capably, if slowly, with every car finishing that day.

While those cars were meandering around DARPA's parking lot challenge, Google started to make its move, hiring as many driverless experts as it could. First, it bought VueTool, a digital mapping technology out of Stanford's AI lab that formed the basis of Google Street View, and a subsequent digital mapping project called Ground Truth that was staffed by Thrun – who gave up tenure at Stanford for the gig, no small thing – Levandowski, Stanford's Mike Montemerlo (who like Thrun also used to work with Whittaker at CMU) and Dmitri Dolgov.

While the team worked on mapping, which used a machine-imaging system developed by Levandowski's 510 Systems, Levandowski also had a side project: building a self-driving pizza delivery car for a TV show called *Prototype This!* Google didn't want to participate in the project, so Levandowski founded his own company, Anthony's Robots, to make what he dubbed the Pribot, using a Toyota Prius and the driving technology developed by 510 Systems. Followed by police and a truck carrying the show's camera crew, the Prius drove along the waterfront and over the Bay Bridge, where it delivered its pizza. Levandowski managed that in part using a detailed map of the area, but also a lidar system atop the car and interpreted by 510 Systems software; the only fault in the 25-minute ride was cutting the turn off the bridge so closely the car scraped its side. It was a major accomplishment, achieved a year before Google began its driverless work in earnest.

Back at Google, the team mapping data quietly started 'Project Chauffeur' at Google's new X labs, an experimental division set up in 2009 to explore 'moonshot' technologies – Google's name for projects that seem impossible but could have disruptive results. Seven driverless cars – again, Priuses – were road-tested around the Bay Area, with a massive bit of kit plonked on the top, not unlike Street View cameras. 'If people asked us what was on the cars, we'd say "it's a laser" and just drive off,' Levandowski said in an interview.*

The great tech journalist John Markoff revealed the project's existence in the *New York Times*, describing a half-hour ride in one of the cars. He described how the car merged onto Highway 101 and managed city traffic in Mountain View via sensors and GPS, though engineer Chris Urmson – poached from the CMU DARPA team – was at the wheel as a safety driver, and had to take control twice in the short journey after a cyclist ran a red light and a car in front backed up to access a parking space.

At the time, Google said in a blog post attributed to Thrun that the seven cars had driven as far as 1,600km (1,000 miles) without any human intervention, navigating the route between its Mountain View campus and its Santa Monica office, as well as taking in tourist sights such as crooked Lombard Street and the Golden Gate Bridge. The

* You can read the Levandowski interview in the *Guardian* here: https://biturl.top/iuqMjm

cars travelled 225,000km (140,000 miles) with only a single accident, when one of them was rear-ended while stopped at a light.

Google heralded an automated future where driver's licences, drunk driving and road deaths would all be engineered out of existence within a few years. All of that was to be accomplished using cameras, radar and that laser range finder that Velodyne had developed – the Hall brothers' team may not have won, but they hit the big time because of the race. And they weren't alone, with developers like Thrun, Levandowski, Urmson and others snapped up by Google after the DARPA races.

Google's work sparked a new race to develop driverless technologies — and win regulatory approval.

Generally, there are two routes to driverless technology. First, we could slowly add automation that, through updates and iterations, becomes more and more capable. This is the Tesla model, where the cars can already manage some motorway driving but require drivers to be paying attention and ready to take the wheel. That doesn't mean they do: in 2018, a report showed that the Autopilot system warned a driver three times to take the wheel before a fatal crash, but the driver never did. Other times, the car fails to spot a problem and therefore doesn't ask for help, including one incident where Autopilot didn't register the flashing lights of a school bus and struck a child (he survived).*

The other road to driverless is to develop cars that are entirely driverless. In 2014, Google revealed its prototype 'Firefly'. The car's key features included having no steering wheel, no brake pedals and one big emergency stop button. The project was later renamed Waymo, and in 2016 the company sent out an engineer's friend for an unescorted ride

* According to *The Washington Post*, as of 2023, there were 17 fatalities and 736 accidents involving Teslas believed to be in Autopilot mode, though causes varied: https://biturl.top/2qU3eq

without a safety driver. That rider, Steve Mahan, happened to be legally blind, helping Google make its argument for the social merit of the project. Either way, by Waymo's reckoning, Mahan was the first passenger of a driverless car on a public road.*

Other companies quickly followed suit. One was Uber, and that's where this story takes a dramatic turn. Levandowski left Waymo in 2016, after earning $127 million throughout the years, to found his own startup, the autonomous truck company Otto.

Otto leapt into the public eye via a test drive in Nevada, for which it wasn't actually licensed, before delivering a shipment of beer – and then being acquired by Uber for $680 million† to boost the ride-hailing app maker's driverless efforts. A year later, Google/Waymo sued Uber and Levandowski for conspiracy to steal trade secrets. Before he departed, Google says, Levandowski downloaded 14,000 files, though Uber disputed any of those files were used in its own technology.

After four days in court – during which Levandowski mostly pleaded the fifth, refusing to talk so he didn't incriminate himself – the two companies agreed to settle, with Uber handing over 0.34 per cent of its stock to Google, at the time worth about $245 million.

That wasn't the end. Levandowski was let go from Uber and that sparked a further legal battle, which led to Uber having to pay him millions. And then came the criminal case, where Levandowski faced dozens of counts of theft and

---

* In the video, Mahan visits a drive-through, though Thrun admits this was staged. After all, just because a car can get a blind person to their destination, how does the person know where to get out, which direction to walk after arrival, or where to grab their fast food order? That aside, visitors to the Computer History Museum in Mountain View, California can sit in a Firefly in the cafe.

† During legal proceedings, it emerged that the actual figure was in fact much lower, at around $220 million, due to the way in which the deal was structured around shares and cash bonuses.

attempted theft of trade secrets. A plea deal saw him admit to just one charge and sentenced to 18 months in prison. The judge in the case, US District Judge William Alsup, said it was the 'biggest trade secret crime I have ever seen'. Sentenced during the COVID-19 pandemic, Levandowski was allowed to report to prison once the public health crisis was under control, and then a pardon from Donald Trump in the final days of his presidency meant he never served any time.

Money is still being thrown at driverless cars, as well as plenty of engineering research and development from people with real experience in automation. Alongside Alphabet's Waymo*, there's rival startup Cruise, which was bought by General Motors (GM) in 2016 for more than a billion dollars, and Tesla has been adding functionality to its Autopilot mode, too. Those are perhaps the leaders, in terms of miles driven and attention received in the media, though there are plenty of others – many of which can be traced back through Google and Uber to the DARPA competition. Ford and VW-backed Argo AI was founded by one-time Google staffer Bryan Salesky, who worked on the CMU driverless car project with Urmson (formerly of Google and CMU), alongside the former head engineer of Uber's self-driving car unit, Peter Rander; it's raised more than $3.6 billion, mainly from carmakers. There's also truck-focused Aurora, which was started by Urmson, Tesla's former head of Autopilot Sterling Anderson, and Uber's former head of autonomy Drew Bagnell – so you can see why that startup has won billions in investment, as well as media attention.

★ ★ ★

* In 2015, Google was reorganised under an umbrella company called Alphabet. So Google and Waymo both became companies owned by Alphabet, meaning journalists had to start littering their stories with explainer phrases such as 'Alphabet, the company that owns Google', and so on.

The Jaguar i-Pace's driving is smooth, but watching the steering wheel turn without human hands makes me feel sick – it's like sitting next to a ghost. I am in the front passenger seat of a Waymo-controlled car. There is no safety driver next to me, so we're entirely trusting the systems developed by Thrun and the engineers that followed him. And 'we' means me, my husband and our 18-month-old daughter in the back seat.

We're staying at a Palo Alto motel along El Camino Real, the road that cuts through Silicon Valley, but Waymo's cars don't operate that far south; they're only allowed to pick up riders in a small chunk of San Francisco. So we've driven our rental car to the Stonestown Galleria, one of the southernmost points that Waymo covers which has an excess of parking – RVs line the roads at nearby San Francisco State University, presumably offering temporary housing in a city with overpriced real estate.

We park up, tap the button in the Waymo app and soon a Jaguar threads its way through the lot towards us. There is no mistaking this as a regular car: there are sensors and cameras on the front and rear, as well as perched atop the roof – and LED lights spell out my initials and say 'hello'. We strap in my daughter's car seat, leaving the doors wide open after the sudden fear strikes of the Waymo driving off with just her. (It'd make for a good anecdote in the book, my husband notes). But the system is too intelligent for that and it waits for us all to enter, strap in, close the doors and confirm our destination on the in-car display. The whole experience is slick and reassuring.

We pick a random set of sights to visit – the Painted Ladies of Alamo Square Park, Golden Gate Park, but not the famous bridge of the same name as it's out of range – to run the computer-controlled Jaguar through its paces. It starts to drive out of the parking lot, slowly and carefully, with a prerecorded voice giving us a safety briefing. 'Oh, I don't feel good,' I say, watching the wheel turn on its own. My stomach churns.

I get over it. Sometimes the Waymo driver makes decisions we wouldn't, clearly erring on the side of caution. It doesn't make a dual turn alongside another car, preferring to wait for traffic to clear further; my husband (an excellent driver) says the gap was acceptable, and he would have gone. On a right-hand turn at a red, the car in the next lane hasn't left much space for us to squeeze through; the car patiently waits instead of inching forward as we would have done. But then the light turns green and it pulls forward like space wasn't the issue. We can't unpick the decision-making: did it hold position because of the light, the squeeze or a combination of both? The car later refuses to pull up next to a large lorry until it can fully pass it; mocking it as a nervous driver, we can't figure out the car's problem. The Waymo later slows down for an object we can't spot, but guess may have been streetcar rails. None of these are major mistakes, though it does make a few more serious blunders, including stopping on a pedestrian crossing.

One upside of the smooth ride and careful driving is that our baby falls asleep; we extend the trip, so clearly my nerves have calmed. People spot the lack of driver and wave at us; one driver is so busy taking video he blocks the road, including us.

⋆ ⋆ ⋆

The streets of a handful of US cities aren't the only places you can spot driverless cars, though trials in the UK are focused on more than the car. TRL has been running various trials, not so much of technology but of everything else that surrounds the idea of automobile automation, including how pedestrians and other drivers interact with driverless cars. In one trial in London, its GATEway slow-moving autonomous pod linked North Greenwich tube and bus station with the O2 concert venue. These TRL pods weren't designed for roads but pavements – or, in this case, they run in bike lanes.

Locals were given a chance to ride in the pods and offer feedback about how they felt, particularly about safety, but the project also considered pedestrians (and cyclists) and whether they minded sharing their space with an autonomous machine. Generally, the feedback was positive, although some information was interesting: tinted windows were a no-go, as people liked being able to see out and have the world see in, especially as the rides were shared.

The GATEway pod wasn't a car and it wasn't tested on roads (though, later, TRL did take it out onto local streets to see how the system could interact with other cars). This is because driverless technology shouldn't be the preserve of private cars but should be used to expand mass transit as much as possible. Slow-moving pods such as the one that trundled through Greenwich could be useful last-mile systems to help people unable to walk longer distances or who simply have too much shopping to carry home.

Up north of Edinburgh, driverless buses are ferrying passengers over the famous Forth Bridge. The CAVForth project, a joint effort by Fusion Processing, bus company Stagecoach and local authorities, is putting the usual collection of lidar, radar and cameras on to single-decker buses in order to see if autonomous public transport can help cut costs, believing more efficient driving will reduce fuel use and wear on tyres.

Automated buses make more sense than cars. The routes are well defined, so there are no surprises, and infrastructure can be updated to assist the driverless system. Out at the Forth Bridge, the road lines were repainted to make them easier to see, connected traffic signals installed so the bus knows when a red light is coming up, and extra CCTV added to help keep watch over the route.

On the other hand, you don't really want a bus with no staff on board. For the pilot stage of this trial, a safety driver was at the wheel to take over in case of any problems and also to manage the exit from the depot; he cheerfully held his hands up to prove he wasn't steering while we traversed the bridge. For the foreseeable future, they also have a second member of staff on board, a 'captain' who speaks to the public and helps with tickets. That's a wise move to make this shift to autonomy more palatable, as it's hard to imagine ever getting on a bus that isn't staffed – as a woman, I can't imagine doing that at night.

★ ★ ★

Perhaps the best way to understand how driverless cars work is to see what happens when they fail. The most significant failure to date is unquestionably the crash of Uber's driverless car in March 2018, killing Elaine Herzberg in Tempe, Arizona. The 49-year-old woman was the first pedestrian to be killed by a self-driving car, sparking an investigation by the American National Transport Safety Board (NTSB).*

Herzberg was struck by the car while walking across the road. She wasn't using a designated pedestrian crossing and it was dark out, neither of which should prevent a driverless car from spotting her, as radar and lidar see well without ambient light. The car hit her at almost full speed, braking just a fraction of a second before the collision. A safety driver was at the wheel; she was not looking at the road but watching a video on her phone. The car was not transporting any passengers but testing the system and training the AI.

The NTSB pinned the bulk of the blame for the crash on the safety driver's inattention and a lack of safety culture at Uber. However, its report also laid out in second-by-second detail how the driverless system processed the object that turned out to be Herzberg and why it failed to take evasive measures or brake in good time. And regardless of whether that system was found to be the main fault in the crash or not, it makes for revealing reading for anyone seeking to understand how a modern driverless car thinks.

But first, a quick rundown of how the system in question operated. This particular iteration of Uber's self-driving tech

* The results of the full NTSB investigation are here: https://biturl.top/BNveme

was mounted on a 2017 Volvo XC90. A lidar unit on its roof, designed to watch blind spots close to the car, could see in all directions out to a distance of 100m (110 yards), while long-range radar also had a 360-degree view. Cameras looked forwards and back, as well as along the side of the car, with sensors on the bumper for close-range detection – though these were disabled at the time. GPS was used for positioning. That sounds like a lot of kit, but the previous iteration of Uber's driverless car system had seven lidar sensors, six more than this version.*

All of the data from those sensors and cameras was fed into an onboard automated driving system (ADS) that had to be instantaneously processed to control the car – steering, braking, changing lanes and so on. The ADS could be disabled by the safety driver taking the wheel, braking or slapping down a big red button in the centre console next to the gear stick. The ADS could also turn itself off, first telling the driver to take over if there's a fault. It could also label and track objects it detected, keeping an eye on them as the car moved in case evasive manoeuvres were required.

The Uber ADS could label objects as pedestrians, vehicles, cyclists or 'other' – that's it, that's as much detail as it could handle. The car knew that a pedestrian at a crossing was likely to cross, and a bicycle alongside the car was expected to follow the traffic flow, so it defaulted to those 'goals' as the anticipated route of those objects, although additional trajectory data was also considered. Any object dubbed as 'other' doesn't have such a goal and was therefore regarded as static until the data showed otherwise.

This was part of the problem. In a country where pedestrians are expected to use designated crossings and failure to do so – jaywalking – constitutes an offence, Herzberg was a

---

* Other driverless systems at the time had yet more lidar systems; Waymo's had six while GM's had five, according to Reuters reports: https://biturl.top/JJfe6z

pedestrian outside a crossing, pushing a bike across the road. The system couldn't identify jaywalking pedestrians so it didn't react until it was too late, according to the NTSB report. The report breaks down how the ADS operated from 10 seconds before impact and gives us a revealing glimpse into the inner workings of artificial intelligence.

Travelling at 70km/h (44mph), the car's radar detects Herzberg 5.6 seconds before the crash, classifying her as an 'other' object. That means the ADS initially labels her as static, and therefore not in the future path of the vehicle.

At 5.2 seconds, the lidar first spots her. It doesn't know that the radar has seen her, so has no record of her movement in that split second. The lidar data also leads the system to classify Herzberg as a static, 'other' object, requiring no action from the car.

At 4.2 seconds before impact, the lidar detects her as a vehicle. Because that's a different classification, it doesn't realise she's the same 'object' as the previous detection, so can't link the data points to see any movement. The ADS still believes she isn't moving, even though she's walking directly into the path of the Volvo.

At 3.9 seconds, the lidar again registers her as a vehicle. Now it sees that she's moving, but because the ADS believes her to be a car, it decides she's moving forward in the next lane over because that is what cars do. As such, it's not deemed to be in the way of the vehicle.

Between 3.8 seconds and 2.7 seconds, the system couldn't make up its machine mind, dithering several times between labelling Herzberg as another vehicle in the left-hand lane or as an 'other' object that wasn't moving. Each time the system alternates on classification, it loses the trajectory data. It has no idea that she's moved at all, even though she's walking into the path of the car.

At 2.6 seconds from the crash, the system spots her bike, classifying her as a cyclist. As the label has changed, there's again no tracking data, so it believes she's not moving. A split second later, it decides she is moving, but assumes it's in the direction of the traffic in the next lane over. The car is now travelling at 72km/h (45mph).

At just 1.5 seconds from the point of impact, the car labels Herzberg as an unknown object. Because it's lost the tracking history by changing classification, it once again defaults to the assumption that she's not moving. It now sees that she's partially in the way of the car, responding by making a plan to dodge to the right to avoid a collision.

At 1.2 seconds before the car strikes Herzberg, it relabels her to a bicycle, again losing any data tracking her movement across the road. It sees she's in the car's path, but the previous plan to dodge around her is no longer possible, meaning the car will have to brake. However, to avoid too many false alarms and the car excessively slamming on its brakes, the ADS has an 'action suppression' system. That's simply a one-second pause to reassess the situation before taking action.

After that one-second pause, 0.2 seconds before impact, the car finally hits the brakes, sounding an alert to the safety driver, who takes over the wheel 0.02 seconds after Herzberg is struck by the car at a speed of 62km/h (39mph) before braking fully.

At no point did the system register Herzberg as a pedestrian. It couldn't predict her trajectory and path because it couldn't decide what she was. When it did realise a crash was imminent, it took a second to reconsider what action to take – not for safety reasons but simply to avoid a bumpy ride. Despite the car seeing Herzberg six seconds before it slammed into her, the system's dithering meant it did nothing to avoid her until a split second before it killed her.

Of course, a human-driven car could have reacted just as slowly. Collisions happen constantly because people aren't paying careful attention; distracted driving killed 3,522 people in 2021 in the US alone, according to the National Highway Traffic Safety Administration. But the flaws in this system weren't only mistakes, such as false assumptions about how cyclists and vehicles behave, but were also by design, with delays added to avoid the car from slamming on the brakes too often. Such details matter: without the one-second pause, the car wouldn't be able to drive smoothly; with it, the vehicle has one second less to brake in an accident. These are serious mitigations.

After the crash, Uber was banned from using ADS cars on Arizona roads, and it withdrew its fleet and paused its research. Seven months later, the cars were back on the public streets in Pittsburgh, Pennsylvania, though initially only on a 1.6km (1-mile) loop at a maximum speed of 40km/h (25mph). Among other sensor changes, Uber disabled action suppression and improved object tracking to maintain data even when the classifications change. A simulation of the incident with the changes in place showed that braking would now have started four seconds before impact.

We don't know if that would have saved Herzberg's life, but it sure as heck would have given her a better chance. But we do know that the ADS needed to be better. A car that can drive around Tempe streets successfully in most situations can still be deeply flawed when someone behaves out of the ordinary. This car could be driverless most of the time, but it needed a driver for five seconds – without one, Herzberg died.

The safety driver faced charges of negligent homicide and significant potential jail time, but pleaded guilty to endangerment and was sentenced to three years' probation. Uber has since ended its research into driverless cars, selling its technology to driverless trucking startup Aurora, led by ex-Waymo engineer Chris Urmson – though Uber also kept its fingers in the pie with a $400 million investment.

★ ★ ★

Here's where we get into semantics: does a driverless car have a driver? I have been in truly driverless cars – no one at the wheel – and I've been in those with safety drivers just in case, or to take the wheel for parts of the route the vehicle couldn't manage, or wasn't allowed to by law. But in all cases, the driverless cars (and buses) are operated in controlled, limited conditions. After their trials in San Francisco and other US cities, Waymo and Cruise want their cars to drive everywhere eventually.

And that's my idea of driverless: no driver required. If you need a driver's licence or to pay attention to the road with your hands on the wheel, it's not driverless. That doesn't mean automation features are pointless; they can boost safety and fuel efficiency and even give drivers a break now and then to reduce fatigue, helping to prevent accidents and make driving a bit nicer. However, that is not a driverless car but one with impressive automated functions.

I find that fully automatic level 5 unlikely in my lifetime. Never having to drive again would be awesome, but the challenges are equally awe-inducing. Consider San Francisco, where both Waymo and Cruise operate. State regulators in August 2023 approved these two leaders in driverless technology to offer paid-for taxi services using their automated cars – but not without friction, including pushback from municipal authorities. After delays to the decision and seven hours of discussion, Waymo and Cruise won expanded operating rights, including the former being able to charge for fares and the latter being able to drive during the day.

But days after, a story about Cruise went viral after several of its cars simultaneously broke down, blocking a road near a music festival; the company said 'wireless bandwidth constraints' from local crowds meant remote operators couldn't fix the issue from a distance. Cruise promised to solve the connectivity challenge, but shortly afterwards was faced with another stuck car: one had driven into wet concrete. You can't say they haven't made their mark on the city.

Since then the dispute has hit the streets. Angered by being forced to live in a testing lab, activists have taken to plonking road cones on the bonnets of empty cars, confusing the system and stranding them. And no wonder. In 2022, a Cruise car blocked a fire truck coming from the opposite direction that entered its lane to avoid a double-parked garbage truck; another nearly clipped a streetcar full of baseball fans, with passengers yelling abuse at the car; and yet another was pulled over by police for failing to have its headlights on, but then drove away – not, according to the company, to evade the authorities, but to find a safer spot to stop. The Cruise car was controlled remotely after police contacted the company.

In the months leading up to the regulatory decision, a dog was run over and killed by a Waymo car after darting into the road; the 54 bus was stuck behind a Waymo that was waiting 15 minutes for humans to come and rescue it; and a Cruise car rear-ended a bus. And bus drivers aren't happy, with one saying on Twitter: 'When an autonomous vehicle causes a collision, it wasn't tired, or intoxicated, it didn't get distracted or try to get away with something it knew better than to do. It "believed" it was driving correctly. They don't work as advertised, and they shouldn't be on the road.' In October 2023, regulators pulled Cruise cars from the city's roads after one drove over and stopped on top of a pedestrian knocked by another vehicle into its driving path; luckily, the person wasn't killed.

All this is after years of training and testing in San Francisco, the home labs of Waymo and Cruise, though driverless cars are also operating in other US cities. Imagine what would happen if these robo-taxis were let loose in a more challenging environment, where there's snow, disregard for traffic rules, more dogs running loose or complicated road systems, like Britain's roundabouts. Much of what these companies and their systems learn in San Francisco will be applicable elsewhere, but there will be lessons to learn in every new city. It's safe to say rollout will be slow – and if you live outside of a major US city with supportive authorities, it'll be even slower.

There are different roads we could take when it comes to self-driving cars: sticking with it, mitigating limitations or giving up.

The first is to keep at it, and Waymo and California's state authorities are hell-bent on continuing to push in this direction. But there's a real risk that regulators will lose patience and ban the cars from city roads, and that investors will give up and yank funding for these high-cost trials rather than continuing to spend billions of dollars with little short-term payoff. DARPA has continued its investment in the area, more quietly than with the Grand Challenges, but has also admitted the whole thing has taken much longer than the US military would like. For city-bound vehicles, limited investor flight is already happening: Uber-rival ride-hailing app Lyft sold off its driverless division for $550 million to Toyota rather than burning through more cash. Chris Urmson has predicted it'll be *another three decades* before driverless cars are widely available.

There's another issue with continuing the development of this driverless technology in this way: it might never work, ever. Or it might require another computing boost, new discoveries in AI, or some other technological leap we can't yet foresee – or perhaps nothing will crack it, no matter what we do.

The second option is to mitigate limitations. This is a clever-sounding way of saying we need to work around the problems. All those historical driverless cars had to keep to motorways; their developers knew this. Perhaps modern driverless cars should too. Or, they'll be as Waymo and Cruise cars are now, always requiring connectivity to be assisted by remote operators. Is that driverless or remote-controlled?

There are other options. Inside cities, designated lanes could be given over to driverless vehicles, keeping people and other cars out of their way; this would be a mistake, given the extent to which roads already carve up and divide cities, but the British RRL researchers considered it one option to introduce the technology to roads. Alternatively, we could limit cars rather than people, perhaps restricting driverless technologies to specific routes where we know they will not have difficulties, and requiring passengers to walk to specific pickup points. In Milton Keynes in the UK, a trial of slow-moving, pavement-bound driverless pods ferried less able walkers from transport hubs closer to their homes, helping them get around more easily without clogging roads. And, as we saw earlier, Scotland already has the first driverless bus route. Using this technology for public transport means it can be kept in rapid transit lanes, away from private cars.

And then there's a third option: giving up. It's not as pathetic as it sounds. We can admit that truly level 5, fully driverless cars aren't possible (or at least not in a reasonable time frame) and instead we can integrate the technology into human-driven vehicles for things like safety, comfort and fitting into tight parking spaces. In this scenario, automation keeps cars in their lanes and maintains appropriate speeds, reducing accidents or lessening their severity. Drivers will still need to hold a licence but can let their minds wander when stuck in traffic or on approved stretches of motorway. That's not driverless, but it's not so bad, either.

Of course, we need not choose any single option. Waymo and Cruise can keep on keeping on as long as they'd like, continuing to use remote controls when needed, and in the meantime, carmakers can integrate these ideas into upcoming cars. One day we may have level 5, but until then we have safer cars. Everybody wins. Well, except the people living in the cities used as labs, of course.

However, there are challenges beyond the technical. Driverless cars are inherently surveillance machines, recording everything in their path, raising concerns about privacy amid reports that policing authorities have already requested captured footage. Waymo and Cruise are developing robo-taxis, not consumer cars – you'll have to sign up for their systems to use this technology. Good luck escaping the tentacles of big tech when they run the internet *and* roads. Tesla and other carmakers are offering their not-fully-driverless systems to anyone with the cash, but research suggests that systems with limited automation may risk the safety of other drivers, pedestrians, and frankly, everything else that exists in the world, small dogs included. And are cars even the correct answer to the question? While it's clear the US isn't going to ditch suburban motorway living anytime soon, cities need fewer cars, not to be flooded with driverless ones wandering the streets while they await a passenger.

Given all these questions, why are we (well, they) even building driverless cars? It's an exciting engineering challenge that could pay off big if ever successful, and even if not via convenience and safety features. Those honest answers aside, developer companies generally list three motivations: to increase safety, to reduce traffic and to optimise driving for fuel efficiency.

Let's look at those in reverse. Driverless cars can drive more precisely than us feeble humans, potentially slashing fuel use. Plus, the taxi model means people need not own a vehicle, requiring fewer to be built – though it's unclear how that works with commutes when everyone travels simultaneously. Better investment in public transport could address both of those issues.

That's also true of our second motivation, to reduce traffic. The argument is that driverless cars will communicate with each other to avoid traffic jams – surely the reduction in rubbernecking at accidents and other incidents alone will be significant. But that assumes all – or enough – cars on the road are driverless, and that won't be true for some time, even in the best-case scenario. And driverless or not, cars don't reduce traffic, they *are* traffic.

And then there's safety. And this is unquestionably an honest motivation for many – Thrun lost a friend in a road accident in his youth, so it's no wonder he'd like safer cars. But there are plenty of ways we could make roads safer right now without spending billions on developing a technology that may or may not ever protect a single life. We could reduce speeds, in particular in cities and towns; in London, cutting speeds to 32km/h (20mph) in some areas has reduced serious injuries and deaths by a quarter. Improving public transport, including taxi availability and accessibility, will reduce the number of cars on roads generally, but in particular could lead to less drink-driving. We also need better road design: road humps ensure speed limit compliance; narrower streets and raised crossings cut collisions with pedestrians; and roundabouts reduce accidents at intersections.

Now, these tactics are for governments to consider, not Google. Still, if reducing road deaths is your motivation, there are faster, proven ways to go about it rather than waiting on a technology that may or may not exist in our lifetimes. We can simultaneously build better streets and public transport while Waymo and Cruise carry on developing driverless cars, but one of those options is getting billions in funding, and the other is being largely ignored. Not only are bus budgets being slashed, but buses themselves are being physically blocked by these driverless cars. Talk about a double blow.

And perhaps that's easy to understand given Waymo is headquartered in Alphabet's network of Silicon Valley campuses. There, public transport is all but non-existent. The CalTrain runs infrequently and the bus network is so limited that Stanford University offers its own local (and free) bus system. Indeed, Google, Apple and Facebook have long offered private buses to shuttle their staff from San Francisco and elsewhere in the Bay Area to their campuses. Getting from my Palo Alto motel to the Computer History Museum would have taken 44 minutes by bus and just 14 minutes by car.* The estimated public transport return journey from a meeting at Google's X labs was more than 33 minutes,

* As mentioned, the Computer History Museum has a Firefly on display in its cafe. Levandowski's GhostRider motorcycle is in the collection at the Smithsonian National Museum of American History, and the first Cruise vehicle was donated to the Henry Ford Museum.

two-thirds of which was walking. Given the summer heat, I pulled out my phone and hailed an Uber.

My driver was chatty and charming, an international student with big dreams: working for one of the tech companies he picked me up from, or perhaps founding a startup that would become a household name, just like his idols had. Ferrying around people like me was how he paid his tuition fees and sky-high rent. Given the slow progress in perfecting driverless cars, he should have plenty of time to earn enough to pay his bills before his idols automate his poorly paid job out of existence.

CHAPTER TWO

Artificial Intelligence

People are worried about AI. There have been claims this technology could 'cause significant harm to the world' as it's 'potentially more dangerous than nukes' and that the risk of harm keeps some 'awake at night'.*

Those three quotes aren't from critics, anti-AI activists or regulators. They're from leaders of companies chucking billions at building artificial intelligence systems: OpenAI's CEO Sam Altman, warning US Congress about a product he's making and developing; Elon Musk, who helped set up OpenAI, uses the technology in Tesla, and subsequently founded xAI; and Sundar Pichai, CEO of Google-owner Alphabet, which widely uses AI and has its own Gemini system to rival OpenAI.

This attitude isn't new. 'Once the computers got control, we might never get it back. We would survive at their sufferance. If we're lucky, they might decide to keep us as pets.' This quote is from a 1970 story in *Life* magazine, said by another AI creator, the great Marvin Minsky. Fear of AI is as old as AI itself.

If we're so scared of AI, why do we keep building it? There's many possible reasons, but here's a few: humans lack patience; have an innate inability to work collaboratively; have egos that desire to be first; and fear 'the other guys' gaining a powerful technology first. That leaves us with

---

* These quotes are taken from the following: Altman (https://biturl.top/EZ7nQf); Musk (https://biturl.top/QfUjYr); Pichai (https://biturl.top/IJjuqi). The quote from Minsky can be seen here: https://biturl.top/FnYZzy

company leaders whimpering about the risks of a technology that *they* are racing to build.

There's also the fact that what constitutes AI – and therefore what scares us – shifts from decade to decade.

But AI is consistent in one way: breakthroughs are massively overhyped and invariably followed by an equally massive backlash. And that leads to funding cuts and slows research, a seasonal cadence referred to in the industry as a 'winter'.

There is a good reason for the hype. As writer Daniel Crevier notes in his 1993 book *AI: The Tumultuous History of the Search for Artificial Intelligence*, it's difficult for a new field of study to draw investment away from more established subjects like aerospace or nuclear physics. Early AI researchers were like 'Davids against Goliaths', he wrote — to gain attention and funding they needed to loudly and enthusiastically proclaim the benefits of AI, regardless of their own doubts. That trend of promising capabilities that can't be fulfilled has haunted AI throughout its existence.

★ ★ ★

The history of AI is as complicated as the subject itself. By condensing it into a few threads, we can start to understand the rise and fall of different ideas and types of systems.

We're keen on AI. So keen that we were talking about it well before computers were even built – okay, not everyone was, but genius academics like Alan Turing, who unpick maths problems for fun, were doing so – and well before we'd even decided on the term 'artificial intelligence'. From there, the history of AI is a series of ebbs and flows, with researchers slowly picking away at problems before breakthroughs push development in a new direction.

The first half-century of AI is full of enthusiasm and debate. The field of study was given its name at a conference at Dartmouth College, New Hampshire in the summer of 1956, arranged by luminaries such as John McCarthy and Marvin

Minsky, who dominated its early years. After a decade of the so-called golden age of AI, it had moved out of mathematical models in academic papers into the real world, with systems actually being built. But after decades of work, it became clear that promises of human-like thinking remained many more decades off.

In the mid-1970s, a damning and influential report laid that out in cold, stark language – and funding vanished. Summer was over, and the first AI winter had arrived. Some researchers kept working towards their preferred systems quietly in the background, but this sparked a shift away from talk of artificial general intelligence to machines that do specific tasks. These 'expert systems' of the 1980s could diagnose blood disorders, sort through complex computer specifications for customer orders and even reveal chemical structures.

These rules-based systems were practical and functional, when they worked. However, critics argued the real world isn't run by rules or sense, and human intelligence is reactive, so algorithms that only work in simulated worlds are inherently limited. Expert systems and rules-based models fell out of favour, cooling a once-hot source of investment in the industry.

Throughout the 1990s, AI continued to languish as a complicated mathematical way of getting very little done. Or at least that's what it looked like from the outside; but researchers continued to plug away at the problem, until computing technology finally caught up. This sparked a renaissance in one largely forgotten idea: neural networks. This deep-learning technique dominates the AI you've used recently. And you have used AI recently, as it powers voice assistants like Siri and Alexa, search engines and the rest of the internet, and streaming services such as Netflix.

AI has matured into something useful, there's no question. But there's also still plenty of hype to contend with, exploding to new levels when American company OpenAI chucked

ChatGPT on to the internet, and everyone went bonkers with excitement and/or fear. Does the hype mean another disappointment-tinged winter is looming? Or is a permanent summer of AI finally here? And are we all going to become subservient to machines?

⋆ ⋆ ⋆

It's worth distinguishing between AI and its subsets, as well as Artificial General Intelligence (AGI). Artificial intelligence is simply when computers can take data, understand it well enough and make a decision based on the parameters given; though it's worth noting that the professor who popularised the term wishes he chose 'machine intelligence' instead.

There are then different types of AI – from a philosophical standpoint and a technical one. Understanding the terms helps explain some of the debate around this technology. First off, there's strong AI and weak AI. Strong AI is a system with the sentience to understand what it's doing, like a person (for the most part). Weak AI can complete tasks, but no one believes it to have any level of consciousness about what it's doing. So far, strong AI doesn't exist and may well never do so.

Strong AI is generally the same as artificial *general* intelligence (AGI). However, some people reserve the former term for sentient AI and the latter for AI that can solve any problem rather than being limited to specific tasks – it's the difference between a human driving and a driverless car with distinct vision, analysis and actuation systems, for example. Either way, when tech CEOs natter about existential risk, they are referring to the possibility of superintelligent AGI that could think for itself potentially making a decision that harms humans (with intent or not). Again, this AGI doesn't exist. The recent leap forward in large language models such as OpenAI's ChatCPT has some people – including those tech CEOs at the top of the chapter – fearing we're on the precipice of superintelligent, sentient AI, while others believe it's impossible to build.

Another bit of jargon for you: narrow AI. Such a system is designed or trained for a specific task, such as computer vision, letting it hunt for cancer in a medical image or read a traffic sign. As Cambridge professor Michael Wooldridge notes in his book on the subject, *The Road to Conscious Machines: The Story of AI*, this isn't a widely used term. As he writes, if you were to use the term "narrow AI" at a conference, everyone would know you were an outsider to the field — simply because, at this point, all AI is narrow AI.

AI is the umbrella term for the whole shebang, but underneath that are different types and techniques. For example, machine learning is a subset of AI that uses models to learn how to do a task rather than being specifically programmed. Deep learning is a type of machine learning with many layers and parameters, while neural networks are a technique for achieving this. Symbolic AI uses algorithms to understand ideas rather than just numbers; this technique dominated the field from its origins for many decades. Expert systems are reasoning tools that work through an if-then structure by applying a knowledge base of rules and facts to questions; these were popular in the 1980s as they could be put to good use by businesses.

Each of these systems can be applied in different ways. For example, natural language processing (NLP) has long been a core focus of AI, as it can be used for machine translation – useful for governments and militaries keeping watch on rival nations – but also because it enables other technologies, be it voice assistants or analysing large tranches of unstructured text. Another way AI can be used is computer vision, letting machines understand objects that they 'see'. Both examples are specialist forms of AI, and complex tasks can be achieved by combining them. Robotics might pair a decision-making tool and analytics for movement with NLP and computer vision to allow an android to walk around, talk and see. A driverless car's system may have separate but linked AIs for a multitude of tasks: computer vision to analyse what the sensors pick up; automated decision-making to decide when to stop or speed up; and a system that can translate that into the car's controls. But what if one single AI could do all of that? After all, humans can. *That* would be AGI.

At the moment, as Wooldridge noted, all AI is weak and narrow. You can see why marketing teams at tech companies avoid these terms, but they're not criticisms, merely descriptors. You can do a lot with weak and narrow AI: think of it as 'focused' and 'functional' if you want to flatter. This type of AI can't think nor understand, but it can reason through decisions based on parameters and guidelines set by humans, and can complete some tasks we can't, or perhaps those we could do, but much faster. And harm can come from this narrow, weak AI, too. Much of modern AI is trained on data we feed it, so it's prone to bias and error, and when used for serious decisions like whom to put in jail, where to deploy more hospital beds, or whom to hire, the potential for harm is genuine.

But those CEOs' quotes given at the top of the chapter aren't expressing worry about this type of harm. Instead, those fears reflect future-focused concerns that superintelligent, conscious systems will overtake our own abilities, leaving us at real risk of being usurped as the big-brained ones on this planet.

How could superintelligent AGI hurt us? One thought experiment often used to explain the threat is known as the 'paper clip maximiser problem'. It goes like this: ask an AGI to make loads of paper clips, and it may choose to wipe out humans or otherwise wreak havoc to achieve its goal as efficiently as possible. The paper clip problem was created by Swedish philosopher Nick Bostrom, who is widely considered the 'father of longtermism', a controversial spin-off from effective altruism that values the lives of future people as much as those around today. Considering the future impact of our actions certainly sounds wise, but practically it's a way to ignore challenges facing the world today. That belief essentially means we should deprioritise challenges now in favour of avoiding long-term existential risk very far in the future – because if we go extinct hundreds of years in the future, there will be many, many would-have-been people who won't get to be happy. Longtermism prioritises avoidance of possible risks to potential people in the future over actual risks to people today on the grounds that there may be more of the former. It gives tech leaders a moral reason to invest in AI, space travel and the like while people starve to death from poverty.

Longtermism has found favour with Elon Musk, who reportedly said it was a close match for his own philosophy, and with others in the tech world seeking to defend AGI, but it equally bewilders many. Many critics of the philosophy, myself included, argue that if you were genuinely concerned about long-term risks, why focus on AGI over, say, climate change? AI can't wipe us out if we do it to ourselves first. And if you're worried about the loss of future humans, what about the future generations lost when people die this very day in wars, through starvation or from preventable illness? Do their never-born children, grandchildren and so on not matter? It would seem not. And this isn't a guess: a research assistant at the Future of Humanity Institute wrote that the lives of people in poor countries are worth less than rich ones, which I believe is referred to as 'saying the quiet part out loud'.*

All this goes to show that the future of AGI is obscure and bleak, or perhaps disruptive and exciting, depending on whom you talk to and their motivations. But in the history of

---

* If this doesn't seem quite possible, do read the full story from *Vice* here: https://biturl.top/UZnIna

AI, we can find a bit more fact and a bit less politics and marketing guff.

* * *

'I propose to consider the question, "Can machines think?"' So begins Alan Turing's pioneering paper that seeks to decide the potential for intelligence in computers, invented only a handful of years before, and still massive room-fillers with limited abilities. Turing's tale is remarkable and depressing: an unquestionable genius, he described computers before they were made, worked as a codebreaker during the Second World War, and to all intents and purposes invented the entire field of artificial intelligence in that paper. Though appointed to the Order of the British Empire (OBE) in 1946, such royal attention didn't stop his persecution in 1952 for homosexuality. Instead, because he was gay, this hero of computing science was rewarded with a prosecution for 'gross indecency' and chemically castrated.* He died in 1954 at 41, apparently by suicide.†

Let's look at his achievements a little more closely, starting with his invention of computers. He didn't mean to create anything; he was merely answering a mathematical problem. That in itself was no small thing: in the 1930s he was still a student at Cambridge University, and the challenge he set himself was solving the newly described mathematical problem known as the *Entscheidungsproblem* – which translates to 'decision

---

* Turing has since been pardoned and is celebrated on the British £50 note.

† The nature of Turing's death has been questioned in recent years. Turing died after eating an apple laced with cyanide, but there is debate as to whether it was intentional or accidental. A half-eaten apple was found by his housekeeper next to his body, but he often ate one before bed; plus, he'd been conducting experiments using cyanide to electroplate spoons, as you do. That suggests an accident was possible, argues Professor Jack Copeland, who is one of the directors of the Turing Archive.

problem', but looks much cooler in German. This problem was devised by the German mathematician David Hilbert, who asked whether an algorithm could prove maths is logically consistent. Hilbert was looking to see if there were any unsolvable decision problems, which are essentially maths problems with a yes or no answer, like 'does one plus one equal two'.

Rather than write pages of mathematical proofs, Turing got a bit creative in his solution, inventing – just in his head, mind you, as a theory – the idea of a machine that could solve infinite maths problems. He solved the *Entscheidungsproblem* by suggesting that such a universal computing machine could be tasked with answering a question about itself, such as whether the machine would ever halt if given an inconclusive algorithm. Because the machines could work indefinitely, we'd never know if it would stop – perhaps the algorithm in question merely takes a very long time, beyond our lifespans, or maybe it does halt in 300 years. We would never know the answer either way. That means the question 'does a computing machine halt?' can't be answered with a conclusive yes or no, meaning maths isn't just about decidable computations.

While that gave Hilbert his solution, don't worry if it raises more questions than answers for you – 'huh?' is a valid response. But in answering the *Entscheidungsproblem* in this way, Turing described what we now call Turing machines, which are essentially computers.

That paper, beyond making Turing famous in his field at a remarkably young age, also sparked work in military circles to develop machines like the Electronic Numerical Integrator and Computer (ENIAC), built by John Mauchly and J. Presper Eckert, and perfected by John von Neumann in 1945. This was a mechanical version of what Turing described, and computers continue to be built in this style now.

During the war, Turing famously worked at Bletchley Park, the mansion house at the centre of Britain's wartime codebreaking efforts, where he and colleagues built on Polish efforts to create a machine to decrypt communications scrambled by the German Enigma cypher device so the Allies couldn't read them. The decrypting machine, dubbed the Bombe, was followed by Colossus, which not only helped the British unpick German messages but is seen by some as the first electronic computer.*

After the war, while living in London, Turing started working on a design for a stored-program computer, but his work developing a physical machine was hampered by the Official Secrets Act; he later wrote software for the Manchester Mark 1, one of the earliest such machines. But at the same time, he laid out questions around computing intelligence, proposing what became known as the Turing test: if a human engaged in a text-based conversation with a machine can't tell if it is in fact a machine or another human, then that machine should be considered intelligent.

The Turing test is still used today; though it's widely seen as insufficient now, it laid out a way of thinking about computer intelligence well before such machines existed. Beyond the Turing test, his paper 'Computing Machinery and Intelligence' raised the issue of machine intelligence and whether computers are thinking, as well as laying out the main objections against the idea, and accurately predicting the path of development of AI, including the use of chess to test systems.

Turing wasn't the only one thinking about computer intelligence. The rise of these machines sparked debate about whether computers were thinking or just, well, computing. Celebrated thinker and MIT academic Norbert Wiener argued that there were parallels between the workings of machines and animals, and that intelligence was the outcome of processing information, founding the field of cybernetics, the science of automatic control systems and the study of feedback as maths,

---

* The Bombe can still be seen at the museum at Bletchley Park just north of London, where visitors can also walk through Hut 8 where Turing worked – and where he chained a mug to a radiator to stop people from nicking it.

whether these are biologically embodied or not. Claude Shannon, considered the 'father of information theory', also successfully programmed a computer to play chess, and later came up with a theory of how that system could teach itself how to play better, the latter of which he couldn't yet build because he didn't have access to a computer at the time. Indeed, all of these achievements happened when you wouldn't need all your fingers to count how many computers were in the *world*. When Shannon later did publish a paper on learning to play games, it was the first time the term 'machine learning' was used in print. This was the early years of AI, and the research often required first building the machines to do the work.

⋆ ⋆ ⋆

At this point, the phrase 'artificial intelligence' hadn't yet been coined. All of the above ideas and debates and work congealed into the idea of artificial intelligence thanks in part to John McCarthy.

In 1988, when McCarthy was awarded the Kyoto Prize for his contribution to the field, he said he'd always remembered the idea of using computers to behave intelligently was sparked when he attended a conference in the late 1940s at Caltech – this was HIXON, the Symposium on Cerebral Mechanisms and Behaviour, where von Neumann delivered a talk. But – ever the diligent academic – he went back to fact-check his own memory and found no one had actually talked about using computers in that way. 'I had simply jumped to the conclusion that people were interested in that,' he said in an interview in 2011.*

* This is from an interview with McCarthy conducted by Peter Asaro with Selma Selma Šabanović at the IEEE History Center in the US: https://biturl.top/bYNBBn The IEEE History Center has a collection of more than 800 oral histories in electrical and computer technology here: https://biturl.top/eeyeMr

After that idea entered his head, he ended up at Princeton as a grad student in mathematics, where he went to speak to computing great von Neumann about his ideas, who told him to: 'write it up, write it up'. But McCarthy didn't for several years more, believing the ideas weren't very good, but also that it was impossible to model how the brain learned facts at that time.

In 1955, McCarthy was an assistant professor in mathematics at Dartmouth College, and along with Shannon (then at Bell Labs), Harvard researcher Marvin Minsky and IBM's Nathaniel Rochester came up with the idea to run a summer seminar on the idea of thinking machines at the college. Famously, there was debate about what to call the conference as at this point, artificial intelligence didn't yet have a name. McCarthy and Minsky were well aware that branding mattered. If they just slotted their work under existing terms, such as rival Norbert Wiener's cybernetics, they wouldn't get the credit for founding a new field. And if the name was misunderstood, it would colour the research – a previous call for papers had used 'automata studies', resulting in a stack of articles about machines rather than intelligence. Others suggested 'complex information processing', but that applied too widely. While Shannon believed 'artificial intelligence' or 'machine intelligence' was too flashy, McCarthy 'decided to not fly any false flags' and admit the goal was developing human-level intelligence. They decided upon 'artificial intelligence', though in hindsight McCarthy preferred 'machine intelligence'.

They proposed a two-month-long study of AI for 10 attendees 'on the basis of the conjecture that every aspect of learning or any other feature of intelligence can in principle be so precisely described that a machine can be made to simulate it'. The aim was to understand how machines use language, solve problems and correct themselves, and the proposal said a 'significant advance' was expected in just a few short weeks. Unfortunately, the conference didn't get enough money and people didn't want to spend the whole summer attending AI talks, so McCarthy's optimistic hopes of a breakthrough towards human-level AI went unrealised.

While no one made AGI that summer, the Dartmouth Conference did launch AI as a subject of study, and many of the attendees became luminaries in the field, with the three American AI labs that grew out of it – at Stanford, MIT and CMU – becoming the main centres for the work. Though the conference became more of a drop-in session for like-minded researchers to share their work, what they did share was remarkable. The first AI program was presented by Carnegie Mellon University's Allen Newell and Herbert A. Simon. Called the Logic Theorist, it attempted to build mathematical proofs based on the rules it was given. Other attendees included game theorist John Forbes Nash Jr; Warren McCulloch, who would go on to help develop neural networks; Ray Solomonoff, who presented his work on probabilistic machine learning; and Kenneth Shoulders, who beyond working on AI would also work at SRI with robotics luminary Charles Rosen *and* develop a flying car. Another person listed as attending the conference was simply 'Shoulders' friend' (they didn't keep great records).

The years after that conference saw an explosion in AI. MIT's AI lab was sparked by McCarthy and founded by Minsky with apparent ease – requiring little more than a quick conversation after bumping into the university's head of research in the hallway – but McCarthy quickly returned to Stanford thanks to an offer of a full professorship and encouraged by his own dislike of shovelling snow. There he set up a computing science lab, and then a few years later, in 1965 – with a burst of DARPA funding – bought a PDP-6 computer to start the Stanford Artificial Intelligence Lab (SAIL). CMU and the University of Edinburgh also set up AI labs in the 1960s, drawing in the talent and the funds.

The mid-1960s were a boom time for AI. Among the most famous developments was ELIZA, built by MIT's Joseph

Weizenbaum, which asks open-ended questions to mimic a therapist speaking to a patient, and SHRDLU built by Terry Winograd, a natural language system that could operate and execute using plain English. Newell and Simon took their work further with what they called a General Problem Solver, and McCarthy built a system to play chess.*

However, not everyone agreed that the progress was meaningful. In 1973, the Lighthill report landed with a thump, crushing the hopes and dreams – and funding streams – of many AI researchers. Commissioned by the UK's Science Research Council and officially known as *Artificial Intelligence: A General Survey*, the report, written by mathematician James Lighthill, made for depressing reading. 'Most workers in AI research and in related fields confess to a pronounced feeling of disappointment in what has been achieved in the past 25 years,' the report reads. Ouch. It goes on to say that 'in no part of the field have the discoveries made so far produced the major impact that was then promised'. Double ouch. Lighthill noted that 'enormous sums' had been spent on machine translation in particular with 'very little useful result'. Lighthill broke down the industry into three sections: automation (making tasks happen), central nervous system studies (understanding the brain), and a 'bridge' category that tried to tie the two together, in particular via robots. And that 'bridge' was where he saw the worst failures, though he also argued that perhaps automation was better served by programming than by AI.

AI luminaries disagreed and were willing to do so publicly in a televised debate arranged by the Royal Academy with an audience stuffed with mathematicians and engineers – can

* Creating AI that plays games is still how many models are built and tested, notably AlphaGo, an AI-powered system designed by Google's DeepMind to play the board game Go.

you imagine such a show being made today?* Regardless of the credibility or accuracy of his damning report, give Lighthill some credit: he showed up to this debate, knowing that luminaries Donald Michie, John McCarthy and Richard Gregory, a leading AI researcher and Bristol University experimental psychologist, would be there and that they would disagree with him. Indeed, McCarthy flew over just for the show.

The debate focused on robotics, but that was really a stand-in for AI. Lighthill laid out his argument as one between specific-purpose robots, which were even then seen as a success, and so-called general-purpose robots. But he noted that automation doesn't even require AI, because control engineers can programme systems perfectly well without AI; he gave the example of aircraft autopilot. That was all fair criticism, but then he dropped this: 'Computers have been oversold. Understandably enough, as they are very big business indeed. It's common knowledge that some firms bought computers in the expectation of benefits which failed to materialise.'

Now would have been a great time for a mic drop and saunter off the stage, but instead, Lighthill said that wasn't even the problem he was there to address. Instead, he asked the assembled researchers to explain their longer-term view of computing – meaning AI – 'to avoid the public being seriously misled'.

---

* As I write this, two CEOs of tech behemoths are preparing to duke it out not via academic debate but with actual fists. Facebook founder Mark Zuckerberg and Tesla/SpaceX's Elon Musk have spent months discussing a cage fight. Like, an actual fight, where they would hit each other. The dispute was sparked by Zuckerberg's Meta unveiling a Twitter-style app after Musk bought that social media property. The level of discourse has slid far in 50 years, that's for sure. As entertaining as such a staged fight might be, you can alternatively watch the Lighthill debate here: https://biturl.top/N7nIje

To be clear, Lighthill saw research value in AI. But with computing science on the technical, automation side, and biology on the brain science side, he believed using AI as a bridge between the two was largely pointless as core challenges couldn't be overcome to move out of labs and into the real world.

McCarthy counterargued that AI is its own basic science, and no one was making this claim of AI being a 'bridge' between brain science and computing science *except* Lighthill: 'Was it merely for tactical reasons that you chose to ignore [whether] anyone was even making this contention?' It's a verbal slap. Lighthill came back that it's implicit in the name 'artificial intelligence', explaining the term to the man who coined it. McCarthy, cackling with glee, happily reminded his adversary that he invented the term, sparking audience laughter – though of course we now know that he wished he'd chosen a different name.

Another Lighthill criticism is the hurdle of 'combinatorial explosions' – when Lighthill says this phrase, he throws his arms up in the air for dramatic effect – which refers to how more data, perhaps from systems leaving the lab and entering the real world, leads to an exponential increase in processing required. This means that even a massive increase in computing power will only allow for moderate improvements in the way these systems work. That is, unless they cheat with heuristic guidance, which allows humans to set rules and give feedback. As an example, he points to chess, saying that despite two decades of work, AI is a decent amateur player but hardly a grandmaster.

McCarthy responded to the combinatorial explosion complaint by suggesting the preeminent mathematician only just heard of the problem, though it had been worked on since Turing. In short, *don't you know that we're already on that?* The attack highlights Lighthill as an outsider.

Lighthill remained unconvinced that general-purpose robotics or AI were possible. 'The science fiction writers, possibly others, will try to keep it shimmering, or appearing to shimmer, there on the horizon in front of us. And there's something most of us want to believe is there, so many people may be disappointed to hear it's not.'

AI's fundamental difficulties were not solvable, according to Lighthill. The others disagreed – it is their field, after all – but they argued progress had been made already and further work would solve the rest. This is a core problem of telling the future: everyone is right and everyone is wrong – in time. None of them could imagine what would come next, or more importantly, when. Indeed, Michie and McCarthy had a bet that AI would beat a grandmaster at chess by 1979. They lost that bet, but they weren't wrong – just too early. It did eventually happen, with Deep Blue in 1997, though using a style of AI the researchers defending AI had themselves set aside, paired with processing power beyond their wildest dreams. So they were correct that AI would become more capable, but were themselves going about it the wrong way. On the other hand, Lighthill was 'dubious to a very high degree about predictions of a general purpose robot' – and he was correct, as we still don't have AGI nor the level of AI mastery to make a thinking, do-it-all robot (there's a whole chapter on that later in the book).

Before we mastered chess in the 1990s, AI had a miniature revival in the 1980s as the idea of expert systems emerged. Crevier describes expert systems as being sold as a 'nonhuman idiot savant' that could do anything from diagnose illness to choose the best wine for dinner. These machines took curated knowledge on a specialist subject and applied rules – such as if-then – making them useful for businesses and the like with specific tasks. The very definition of narrow AI, expert systems left behind the philosophical questions about whether machines could think and what was the nature of intelligence in favour of practical systems that could perform useful tasks. For example, computing company DEC used an expert system called R1/XCON to process orders for its complicated systems; the high number of component options meant it had more than 17,500 rules.

* * *

Work on the first expert system began in the 1960s at Stanford University, led by Edward Feigenbaum and Bruce Buchanan. Their Dendral, or dendritic algorithm, was designed as a tool to better understand scientific knowledge, in particular organic chemistry, but was later used as the basis for other expert systems. That includes MYCIN, built by a Stanford student in the 1970s to analyse medical data, look for blood diseases and prescribe antibiotics dosages, all using a set of 600 rules. Because it was designed to be used by doctors, MYCIN revealed how it came to a conclusion, rather than operate as a so-called black box that spits out answers with no explanation. Though it worked, the system was never used outside of the lab, because it took so long to enter all the required patient data.

The success of this idea helped draw investment back into AI, though no one called it that. Instead, companies bought expert systems and the government invested in intelligent knowledge-based systems.

But AI researchers can't help themselves. Expert systems worked, when they did, because they used curated knowledge sets and worked to specific rules to solve specific questions. But then Doug Lenat came along and decided to use the idea to build an AGI known as Cyc, which required all of humanity's common sense to be written down and encoded. That manual labour, Lenat predicted, would take 200 years of effort. Naturally, Cyc never accumulated enough knowledge. But it did show that giving AI some knowledge to work from could help point it in the right direction – an idea shown with heuristics in the early years, which gives AI enough built-in information to measure its own success.*

The extreme example of Cyc highlights the flaws in expert systems: common sense is hard to unpick and encode, and gathering expert knowledge requires the participation of experts, who are bound to see that the system could cost them their jobs. Expert systems continued to be used, but the hype ended with another thud. McCarthy argued expert systems were too limited and lacked context, pointing to an example where MYCIN prescribed two weeks of an antibiotic to treat an intestinal infection. That was the correct drug and the dose would have treated the bacteria, but by the time the two weeks were up, the patient would have been long dead.

So after all the hype and excitement around expert systems, DARPA took a closer look at AI, believed it to be of limited use and slashed funding.† And it wasn't just the Americans who were disappointed with AI. Japan had invested almost half a billion dollars into AI and logic programming between 1982 and 1990, via its Fifth Generation Computer Project. But the funding wasn't renewed.

Attention turned away from AI, but work continued out of the spotlight. And new ideas came to the fore – including the need to bring AI into the real world rather than keep it in the

---

* As Wooldridge notes, heuristics are at the core of the 'knowledge graph' that Google uses to tie together concepts to give its search algorithm a bit of a boost. And, at the time of writing, there was plenty of chatter that OpenAI was adding heuristics to its massive models to step closer to AGI. Whether or not that proves true, it's clear the idea has legs.

† In Nick Bostrom's book *Superintelligence*, he quotes Jacob Schwartz, then head of DARPA's Information Science and Technology Office, as describing the history of artificial intelligence as 'consisting always of very limited success in particular areas, followed immediately by failure to reach the broader goals at which these initial successes seem at first to hint.' However, in the footnotes he explains that Schwartz was 'characterising a sceptical view' rather than necessarily sharing his own, the belief often attached to that quote.

lab. AI was developed and tested in environments dubbed 'Blocks World' – picture a tabletop covered in toys and blocks. This was the basis of plenty of AI claims, though sometimes it wasn't even a real tabletop but a simulation. During the Lighthill debate, for instance, Donald Michie showed a video of a robot assembling bits and pieces of a toy car on a table; the idea was that this could be applied to a warehouse to assemble real cars. But obviously, real cars and factories are more complicated than toys on tabletops. So it was time for AI to leave Blocks World and enter the real world.

Such was the argument made by Rodney Brooks. Own a Roomba? (Or a knockoff?) Then you have Brooks and his argument for real-life AI to thank. Beyond co-founding the robot vacuum company iRobot with Colin Angle and Helen Greiner, Brooks argued intelligence was about understanding the contextual environment and reacting to it rather than just knowledge representation and reasoning. This is called behavioural AI. A Roomba needs to be able to find its way around a space it's never seen before, not just define a carpet.

So AI was starting to move into the real world, developing real-sounding chatbots, turning neural networks to handwriting and speech recognition, and finally beating a grandmaster at chess with IBM's Deep Blue, originally developed at CMU. That was achieved using symbolic AI algorithms running on a supercomputer to search and assess as many possible positions as possible. In 1997, after losing multiple times over 10 years, Deep Blue finally beat grandmaster Garry Kasparov in a six-game match. At the time, Deep Blue was ranked as the 259th most powerful supercomputer in the world – in short, it was brute force that won the match, rather than raw intelligence.

Though Deep Blue helped reignite public love for AI, the return of the boom times had nothing to do with the architecture that powered that chess-playing system – instead, it was time for the rise of neural networks.

Like AI as a wider discipline, the subset of neural networks has had both booms and busts. Before we get into the history, a quick refresher: one technique for achieving AI is machine learning, in which models are fed data to make sense of on their own. Deep learning is a type of machine learning that uses multiple layers of neural networks which mimic brain neurons by using layers of nodes that connect to each other in specific conditions to pass along data to other layers.

Neural networks take input data and multiply it by weights, an acknowledgement that not all data is equal – if the result is deemed relevant, important or significant, it can be passed on to other neurons to assemble together. Incorrect or insufficient data leads to incorrect outputs.

This process of feeding such a system data and tweaking the weights to get them correct is referred to as training. There are two types of training: supervised and unsupervised. With supervised training, the data set is labelled; with unsupervised, it is not. So if you're looking at a million pictures and want the system to pick out the cats, you can train by labelling some photos as cats, showing those photos to the machine and letting it look for other cats, telling it when it's wrong. With unsupervised training, you skip that first step and simply tell the system whether it's wrong or not.

Both of these practices can lead to circumstances in which the system is looking for features you wouldn't expect. One example is a system which was trained to separate dogs and wolves in an image set. Rather than look for what we would consider wolfish characteristics, the system learned to look for animals sitting outside in the snow. We simply don't know what these systems are looking for, and that makes the whole process rather spooky.

Neural networks are as old as AI itself – older, in fact. And they didn't initially have all that much to do with computers. In 1943, a pair of academics – neurophysiologist Warren McCulloch and mathematician Walter Pitts – described how a Turing machine could be built by mimicking neurons, following that up with a paper a few years later with 'nervous net' designs representing what is widely considered to be the first mathematical model of a neural network.*

Donald Hebb, a Canadian psychologist, extended that work in 1949 by suggesting a theory for how neurons interact when the brain learns something: in particular, the idea that one neuron firing to another builds or strengthens the connection between the two. The first real neural network was then developed by Cornell psychologist Frank Rosenblatt in 1958. He decided to keep it simple and based his perceptron model on the decision and response system in the eye of a fly – neural networks really are inspired by nature. He then added the ability to learn weightings by running inputs multiple times until the output matched with expectations.

His perceptron model is a basic version of a neural network, capturing the essence of the idea. Data points are fed in, weighted for importance and combined; if they go over a chosen threshold, they're deemed relevant. For example, to use this model to decide whether or not to go to the beach, we could have three inputs – how crowded is the beach? have there been shark attacks? what are the waves like? – which all have differing weights based on their importance.†

That's the model. Now we need the input data; it's binary, as we're just doing yes or no questions. Today, the waves are good, there are no sharks and the beach is busy. We can plug

* Crevier notes that the idea of feedback loops which appears in Pitts's and McCulloch's work may sound familiar from Wiener's earlier work – and that's for good reason: they were friends and had a club of scientists to talk through their ideas, which also included John von Neumann. They called themselves the Teleological Society, which probably says a lot about who they were as people that they couldn't just hang out and chat, but had to have a name for it. Crevier adds that the group had a British counterpart called the Ratio Club, which Turing would drop in on from time to time.

† This is adapted from an IBM explainer about neural networks, available here: https://biturl.top/6Freae

all that in for our output or predicted outcome, which is a fancy way of saying the answer to our question.

Here's what it looks like in a sentence: 'Should I go to the beach? The waves are good, which is very important; there are no sharks, which is less important; the crowds are big, but that's not very important. Two important points out of three are positive, so, yes, let's hit the surf.'

Here's the thing, though: that model might not be right. Perhaps the quality of the waves should be more important. Or the threshold to take action is too high. Or there are more factors which need to be considered – what if it's raining or the beach is closed or you only surf with a partner and no one is available? None of those points are currently considered, and even if they were, we don't have the data points to fill those inputs. There are various techniques to boost the accuracy of a model in training, including reinforcement learning, which is when a system learns through trial and error; cost functions, which compare the model's outputs with what they should be; and ideas like backpropagation, which involves automatically going back in an algorithm to figure out where mistakes are being made in order to fix them. Inaccuracies can be caused by incorrect weights, misaligned thresholds and simple bad data – if there actually are sharks in the water, that would be a troubling problem to find out once on your surfboard.

That's Rosenblatt's perceptron (though I'm sure he never explained it with surfing or sharks) and he actually managed to build it, too. Indeed, the US Navy worked with Rosenblatt to build an electronic version of the perceptron, using the Weather Bureau's $2 million '704' computer. It learned to tell left from right after 50 tries in a media demonstration, according to a report in the *New York Times*. That's certainly neat, but the Navy spokesmen told the assembled press that one day an electronic computer would be able to walk, talk, see, write and even be aware of itself, with Rosenblatt saying the machines would be the first to think as humans do and could be fired into space as 'mechanical explorers'. These are big plans to imagine for a machine that marked punch cards as left or right.

One of the problems with Rosenblatt's model was that it only had a single layer. Over at Stanford, two researchers, Bernard Widrow and Marcian Hoff, developed a pair of neural network models. The first was ADALINE (Adaptive Linear Element), which recognised patterns in binary code and predicted what value would come next. That was expanded with MADALINE (Multiple Adaptive Linear Elements), which was developed to reduce echoes on phone lines – it worked. Indeed, it worked so well, the system remains in use today.

Instead of having one neuron, as in our surfing example above, in a neural network there would be multiple layers of neurons, with the inputs given different weights and mixed and matched in different ways. That complexity requires error checking technology, which Rosenblatt successfully developed. However, these early neural nets were limited in what they could achieve, and that lack of utility was laid out in great detail by Minsky and Seymour Papert in a book called *Perceptrons* in 1969.

Widely seen as a call to arms against perceptrons, the book damned neural networks to the background for years. But that might not have been the authors' intent. Indeed, though they methodically pointed out flaws in his work, they dedicated the book to Rosenblatt, who was a childhood friend of Minsky. Regardless of the intent, by the end of the 1960s, neural nets were largely out of favour. Researchers in this field called Minsky 'the devil' for his impact, as AI research was subsequently dominated by the Dartmouth researchers' favoured symbolic AI; Rosenblatt missed the chance to prove them wrong as he died in 1971. This led to neural networks having their own 'winter' just before the wider field fell fallow. Neural networks were clearly always ahead of their time.

While one book perhaps heralded the short-term demise of neural networks, another marked their return 17 years later: *Parallel Distributed Processing* by David E. Rumelhart and James L. McClelland. In 1986 the authors argued that the reason the brain is so good at thinking is because neurons all work at the same time, rather than sequentially (one after the other) as previous neural network models attempted.

This would require layers of neural networks. You might be wondering why it took decades to figure out that multi-layer networks would solve the problems of single-layer networks, but the simple fact is that nobody knew how to build them. In particular, no one could figure out how to train such a complicated system, or how to adjust the weights.* Finding the answer was what made neural networks possible.

Backpropagation works from the output back to the input, adjusting the weights as it goes, with the aim of shrinking the difference between what the network initially outputted and a defined desired output. It's unclear who first originated this idea, though it may well have been invented at multiple different times. In 1974, Paul Werbos published his dissertation explaining how to use backpropagation of errors to train neural networks, though his work was little noticed until 1986 when it was cited in a paper by Rumelhart, Ronald Williams and Geoffrey Hinton, the latter now considered one of the 'fathers of deep learning'.

So in the mid-1980s, multiple researchers came to the same solution for neural networks. But you may have noticed that the mid-80s is not when neural networks took off. There's a

* This remains a problem even today. And if you meddle with the weights or tweak other aspects of a model, it will spit out absolute nonsense. Researcher Janelle Shane does this for fun, spinning the dials on AI systems for comedic effect to suggest paint names like 'Turdly' and new planets called 'Tina'. Give an AI no randomness and it just repeats the most obvious words; give it too much, it's garbage; but when just right, it's hilarious. Her website is here: https://www.aiweirdness.com/

good reason for this: computers couldn't handle them. Neural networks give better outputs the larger and more complex they are, and that requires more and more data for training. A neural network in 1990 would likely have had about 100 neurons – a massive leap from Rosenblatt's work – but 15 years later that would have leapt to a million.

For neural networks to succeed, three problems needed to be solved. First, new algorithms and techniques were required to connect and train more and more neurons – let's call this the maths problem. In order to feed that system, we needed data; and in order to process it all, we needed more powerful computers. In short, we needed researchers to build better models, huge data sets to train those models and powerful computers to run it all.

Deep learning dates from the mid-2000s. A type of machine learning, it includes layered or complex architectures – think of those input-neuron-output algorithms, and then add many layers, make them non-linear, and with a hierarchy for multiple stages of data processing. Models can be applied to different applications, and aren't task-specific, depending on how they're trained. That can be done in many ways: supervised ('This is a cat; what's that?' 'A cat.' 'Well done!'), unsupervised ('Is this a cat?' 'No.' 'Well done!'), or part-supervised ('Here are some cats. Here are some unlabelled pictures. Are they of cats?').

Deep learning is powered by neural networks, and there's a variety of different types, including convolutional neural networks, which analyse data such as images by starting at very basic features such as brightness before zooming in to higher resolution; and recurrent neural networks, which have memory so can go back and examine earlier pieces of data for better context, making them helpful for language and so on. Different algorithm types are useful for different tasks.

Developing these systems and figuring out backpropagation is one part of the 'maths problem', and Hinton and other researchers carried on toiling away at deep learning when everyone else got distracted – he and the rest have been repaid marvellously for such dedication, as is only right.

* * *

Another problem is data, which was initially solved by Fei-Fei Li, director of the Stanford Artificial Intelligence Lab and co-founder of the Stanford Institute for Human-Centred Artificial Intelligence (HAI). In the mid-1980s, a team from Princeton made an English language database called WordNet, enabling researchers to develop natural language processing, systems that can read and write. Aware of the work happening to develop models, Li wanted to do the same for images, so set up ImageNet.

This wasn't just a big pile of images. Each needed to be carefully labelled – this is a cat, it is orange, and so on. Initially she intended to pay undergrad students to label the images, but realised it would take a century or so too long. She tried an algorithm instead, but it lacked accuracy. A grad student supplied an answer: Amazon's Mechanical Turk. This freelance work platform meant she could scale up the work by hiring huge numbers of people to do small amounts of work each. After two years, ImageNet had 3.2 million images labelled and categorised; there are now 14 million.

That gave researchers a data set comprising objects, scenes and images on which to train AI, letting them test their systems – but they also noticed that the data could help improve their systems. A researcher could use ImageNet to first train their model and then improve it, before flipping it over to another recognition task.

Li followed up one excellent idea with another: a competition. The first ImageNet Large Scale Visual Recognition Challenge was held in 2010 but it wasn't until 2012 that everything changed. That year, the winning team was the first to post an accuracy rate above 75 per cent and did it using deep learning

– the team was made up of Geoffrey Hinton, Ilya Sutskever and Alex Krizhevsky (note all of those names).

The next year, the competition was won by one of Hinton's students, Matthew Zeiler, and in 2014 not only was the winner using deep neural networks, but so were *all* the high scorers of the challenge; specifically, these were convolutional neural networks. (These were first developed by Yann LeCun, then of Bell Laboratories, though he called them 'convolutional networks', because neural networks were still out of favour.) By the time the final competition had run in 2017, the accuracy rate of winners had climbed from 72 per cent to more than 97 per cent.

Hinton and his students immediately flipped their project into a company called DNNresearch that had nothing but a name – no products, no proposals, just their names and a website that said nothing. Hinton and his students were being courted by big names in the industry, and he figured selling a startup run by the trio would earn more than salaries if they got hired. A reverse auction for DNNresearch saw four companies bidding: Baidu, Google, Microsoft and DeepMind (which was later bought by Google). Google won the auction with a $44 million bid.*

The first two problems of neural networks were now solved: the algorithms (with Hinton, LeCun and their colleague, Yoshua Bengio, winning a Turing award for the work), and the data (Li and ImageNet). So that left just computing power. Though improvements had been made in hardware performance, processing times and costs were holding back this technique.

Where academics failed to find a solution, gamers had one ready-made. Computing has long had a cadence – a tick and a

* As told in Cade Metz's book *Genius Makers*, the auction rules gave the companies an hour after each bid to raise it by another million; if no one did after that time, the deal was done. This carried on until the price topped $44 million, with only Baidu and Google left. Hinton called it, deciding he'd rather work for Google. For that price, Google got one heck of a deal.

tock. Chip giant Intel would steadily release a faster chip every 18 months or so, and follow that with efficiency improvements. That meant the computing industry got a performance boost followed by cost reductions at a steady rate, largely following Moore's law – named after Intel founder Gordon Moore, who predicted that the number of transistors in chips would double every two years, making them more powerful and more efficient. But that rate of improvement simply wasn't enough for the massive neural networks that companies like Google wanted to build, and consequently they were starting to design their own hardware, data centres and servers. That meant they were open to solutions from outside the norm.

In 2010, two former Stanford colleagues met up for breakfast. One was Bill Dally, working at graphics hardware developer Nvidia, while the other was Andrew Ng, working on a project at Google's X labs to build a neural network that could learn on its own. He was teaching it to find pictures of cats on the internet.

And that project required thousands of CPUs, aka central processing units, which power computing. Nvidia didn't make those; instead it made GPUs, or graphics processing units, which were designed to handle specific and intense workloads such as graphics rendering. Dally figured he could help Ng with just a handful of GPUs. In the end, it took 12 to replace those thousands of CPUs, with the ensuing shift to GPUs sparking an AI boom. Research got cheaper and easier, and Nvidia now dominates the market for AI chips.*

★ ★ ★

* Dally's colleague, Bryan Catanzaro, asked me to set the record straight. There's an assumption that Nvidia lucked out and GPUs just randomly happened to be good for AI. Nvidia had long known its chips were ideal for AI, and he'd been working on developing GPUs for that very purpose since he was a student at Berkeley, well before Ng and Dally caught up over breakfast.

The summer of AI had returned, and it was all about neural networks. Six months after buying Hinton's DNNresearch, Google unveiled a new tool at its developer conference: the ability to search your own photos using image recognition. That's how quickly Google put deep learning to work. 'We took cutting edge research straight out of an academic research lab and launched it, in just a little over six months,' wrote Google's head of the image search team in a blog post.

A year later, in 2014, Google bought DeepMind, a medical AI company, for £400 million, a massive sum of money for another company that hadn't yet released a single product. As with DNNresearch, the aim here was to capture the people – in this case, Demis Hassabis, Mustafa Suleyman, and Shane Legg. Two years later, DeepMind's AlphaGo algorithm beat the best human players. Around the same time, Ian Goodfellow, then at Google, came up with the generative adversarial network (GAN), where two different types of neural networks play a game: one makes output that looks real, while the other has to guess what data is made up, creating a positive feedback loop that makes better and better images (or other output) – yes, deep learning is weird.

In short, there was suddenly tons of money floating around for anything deep learning related – the amount of VC money doled out to pure AI startups leapt by 30 times from 2010 to 2014 when it topped $300 million – and breakthroughs were coming thick and fast.

At the core of machine learning is training. The AI systems are trained on sets of data – not every piece of data in the world, as that would be difficult and pointless – but we give the AI enough pictures of cats, it learns what a cat should look like, and then we show it a photo and it can tell us if there's a cat or not. For that we need a lot of pictures of cats, and many sorts of cats, otherwise the system will only think that cats are orange or fluffy when they can also be calico and short-haired.

Reinforcement learning gives the system a reward for getting a decision right, though it's difficult to teach the system which bits it got right and which it didn't, as it may have made several decisions to create the final output. This results in it not knowing which individual instances were incorrect. It's sort of like training a dog, how it's important to give instant feedback rather than wait even a few moments, as the pooch might misinterpret its activities in the 10 seconds after rolling over as the reason for the treat, rather than the trick itself.

However bias sneaks into data sets in many ways. Humans need to make decisions on what data points to include, as a system with too many pieces or types of data will be slow, and leaving some out (or including others) introduces bias. Image sets, for example, may not have enough examples of people with black skin, meaning a system has fewer examples and is less good for those people.

Naturally, once AI started to be used in the real world for tasks more serious than spotting pictures of cats online, a mini backlash began. No one was demanding the removal of all AI, but there began to be calls for a better understanding of 'black box' systems such as AI-powered algorithms, especially when their results have serious implications for humans.

AI is useful, there's no question. Like any tool, it can be used well or incompetently, or be the wrong tool, or be used (intentionally or not) for outright evil. Take a hammer: perfect for smacking in a nail. You could also use it to dig holes in soil for seeds, though you'd do better with other tools; and it's of no use with screws and even less use to decorate a cake; but if you want to break a window to rob a house, a hammer will do the trick just fine.

This is obviously a very childish way of looking at the world of technology – hammers don't make their own decisions, or impact millions of people – though it's a metaphor that's trotted out with some frequency. But it's worth noting that AI can do good work when applied appropriately and intelligently for relevant tasks; that the wrong type of AI, or even AI overall, simply can't answer all questions or fulfil all tasks; and that when used maliciously or even just unthinkingly it can cause real harm, and even kill someone.

There are plenty of ways AI is used well, to good purpose, without much risk of harm if attention is paid – we're talking hammer to nail, mind your thumb, here. Modern computer games couldn't exist without AI to power the hyper-realistic images or reactive game play. Amazon's Alexa and Apple's Siri are both AI systems, and the reason smartphone images keep getting better is AI-driven processing. In healthcare, applied with ethics and caution, AI can save the lives of patients and ease the workloads of medical technicians, with deep learning taught to read eye scans, examine mammogram images and spot neck cancer. Alaskan wildlife researchers used facial recognition developed for a dog photo filter app to identify and track bears. Conservationists from the Zoological Society of London turned to AI to automatically analyse thousands of hours of sound recordings to pinpoint British animals as part of a biodiversity project, while deep learning models were trained to count colonies of seabirds and follow whales by their songs. AI has been tasked with everything from modelling the impact of climate change to fighting parking tickets and even unlocking protein structures, a long-running challenge in biology.

We don't want to lose any of those helpful benefits. But it's easy to see why people are so against AI when there are so many well-intentioned, poorly implemented instances – sometimes using AI to solve a problem *is* like using a hammer to ice a cake. Amazon trialled the use of AI to sift through CVs, and the system methodically stacked most of the women in the 'not this time' pile. ProPublica revealed in 2016 that software used to predict recidivism was biased against Black people, meaning they are recommended to be held in jail while white people with similar situations were let out on bail. Facial recognition tools can help policing agencies find criminals by comparing CCTV footage to image databases of previous criminals; this might be a time and budget saver, but incorrect arrests have led to innocent, predominantly Black, men spending days in jail. Some of these problems could in theory be solved by better-quality AI trained on unbiased, accurate data sets, but while we experiment with these tools, real humans are being hurt – and more often than not they're from low-income or minority demographic groups.

And then there are the malicious uses of AI – picking up that hammer to smash windows or skulls. Hackers are turning to generative tools to churn out malware and spam that's personalised, massive in scale, or simply better quality, all at lower cost. AI-powered image generation and editing tools have contributed to the rise in deepfakes – everything from the Pope in a funny jacket to the Ukrainian president announcing a surrender in the war with Russia – and revenge porn, with ex-partners' faces superimposed on to porn stars' bodies. Natural language generators can help churn out misinformation and disinformation, making it easier to sow confusion ahead of an election or flood the internet with so much nonsense that it's impossible to understand what's true anymore.

The most bewildering threats are known as adversarial examples. Neural networks don't know what a cat is, they learn to identify a cat through characteristics that may make little sense to humans. By unpicking that, it's possible to figure out how to manipulate them: in one example researchers meddled with images to make a picture of a turtle appear, to an AI system, to look like a rifle. Others have tricked driverless car systems into failing to halt at stop signs with the careful placement of a few stickers that tricked the AI vision system into misunderstanding what it saw.

And all of this is before we get into the so-called existential risks of AI, in which machine superintelligence overtakes our own and we can no longer keep up.

Figuring out how to use AI to benefit humanity while regulating misuse is no easy task, but it's one that's often superseded by that idea of existential risks. It's hard for politicians to focus on one Black man incorrectly arrested or someone losing housing benefits or not getting a job at

Amazon when billionaire tech geniuses are screeching about mass job losses and even the risk of human life being entirely extinguished by these systems. Though it's fair enough to find both unsettling, only one set of problems is happening right now and it's not the dramatic hypothetical.

⋆ ⋆ ⋆

Much of the debate as I write this book has been spurred by the rise of a very specific type of AI: large language models (LLMs). These have hundreds of millions of parameters and chew up huge data sets – such as the entire internet – in order to be able to respond to our queries with natural-sounding language. Examples include OpenAI's GPT, Google's Bard and its successor Gemini, and Facebook's Llama.

OpenAI was founded in 2015 – one of its co-founders being Hinton's former student and DNNresearch co-founder Ilya Sutskever. Initially set up as a non-profit, it was partly funded by Elon Musk, who had a board seat before he ditched in 2018, as well as Reid Hoffman and Peter Thiel. Shortly after, OpenAI shifted away from a non-profit to a capped-profit model, allowing it to take on investment, and after operating relatively quietly for a few years, the company made an announcement: its GPT model was so brilliant – and therefore so terrifying – that it would only release a tiny bit of its capability, so it couldn't be used by dark forces to meddle with society. Or as the company put it: 'Due to our concerns about malicious applications of the technology, we are not releasing the trained model.' Journalists – myself included – cheerfully had a go with the tool, mocking its mistakes and pointing out its inabilities.*

---

* In an article I wrote on language generation from 2020, I asked such AI to write the Queen's Speech, a traditional address in the UK. I will never not laugh at this: 'In the words of our poet laureate: Coventry is right.' http://bit.ly/3tdUD7W

Two years later, the laughing stopped. OpenAI unveiled ChatGPT in November 2022 and journalists – inside tech and in the mainstream media – were astonished by its capabilities, despite it being a hastily thrown together chatbot using an updated version of the company's GPT-3 model, which was shortly to be surpassed by GPT-4. The ChatGPT bot sparked a wave of debate about its impact and concern regarding mistakes in its answers, but also excitement, with 30 million users in its first two months. It has also triggered an arms race: Google quickly released its own large language model, Bard, and though it too returned false results – as *New Scientist* noted, it shared a factual error about who took the first pictures of a planet outside our own solar system – it was unquestionably impressive. OpenAI responded with GPT-4, an even more advanced version, and Microsoft chucked billions at the company to bring its capabilities to its always-a-bridesmaid search engine, Bing.

In response, thousands of industry leaders – including Musk – called in an open letter for AI labs to take a six-month breather on the development of systems more powerful than GPT-4, a very specific milestone that ensures OpenAI keeps its lead. Shared on the website of the Future of Life Institute we discussed at the top of this chapter, the letter does warn about harms happening now – in particular disinformation and the risk to jobs – but focuses largely on concerns about non-human minds replacing us and taking control of our civilisation. 'Powerful AI systems should be developed only once we are confident that their effects will be positive and their risks will be manageable,' the letter says.

Geoffrey Hinton didn't sign that letter, but the so-called godfather of neural networks did something even more unexpected: he stepped down from his role at Google so he could more openly discuss the threats. To be clear, he didn't quit because he was worried about Google in particular, but about the wider industry. Hinton's warnings were twofold. One, he was concerned about how AI was being used in the here and now, for disinformation and the like. And two, he thought we ought to prepare for the consequences of when superintelligent AI arrived, which was now looming sooner than he'd always thought.

He rationalised his work, he told the newspaper, with the excuse that if he hadn't developed these technologies, someone else would have — and that it's difficult to prevent people from misusing AI to bad ends. His argument sounded reasonable among the shrill crowds, and given he actually understood the systems in question, was a welcome voice of concern.

At the time of writing, AGI, or super-intelligent AI or strong AI – whatever you want to call it – doesn't exist. It might never exist, or it might have been created before this book is published. But there are a few things to note. One, LLMs aren't AGI. They are good-ish at writing, and will presumably get better. But what they do is put one letter after another, based on a trawl of the internet and Wikipedia and other bodies of words.

They have no sense of what is true, so often write text that sounds good but isn't accurate. That removal of meaning from language is worth noting; the whole point of writing is to communicate meaning, after all. People have referred to these inaccuracies as glitches or 'hallucinations', completely misunderstanding what LLMs do: they generate language, not truth. Despite this, some believe these systems are on the verge of becoming sentient and overtaking our capabilities of understanding, but it's never really made clear how a system goes from putting one word after another to having a sense of self with wants and desires and the ability to take action.

Some argue LLMs are learning common sense: researchers at Microsoft, which as noted above has invested billions into OpenAI, published a paper arguing that GPT-4 is capable of reasoning and common sense, showing a real step towards AGI. To show that it's learning rather than just regurgitating, they asked it to draw a unicorn multiple times over the span of a month, subsequently reporting that the images became more sophisticated. It's hard to take people seriously when they're arguing that they've spotted the seeds of nascent AGI in a frankly terrible vector drawing of a mythical creature.*

With regards to common sense, the researchers argued it can answer questions that require a basic understanding of how the world works – things like gravity, directions and so on. For this paper, the researchers used a classic example: 'a hunter walks one mile south, one mile east, and one mile north and ends up right back where he started. He sees a bear and shoots it. What colour is the bear?'

The more limited ChatGPT declares the question unanswerable because no data is given about the colour of the bear. This is correct, and it's a perfectly fine answer: for all the machine knows, before the hunter wasted his time taking this one-mile-each-direction stroll, he may have spray-painted a bear pink.

But according to the Microsoft researchers GPT-4, the more advanced version of the OpenAI system, methodically works out where the hunter is located, because the only place where that walk would bring you back to the same point is at the north pole, where the only bears are polar bears, and therefore white. (Aside from the one I just painted pink.)

Here's the problem: GPT-4 doesn't know if I've painted a bear pink, or if I'm referring to violence against a teddy bear, or something else equally weird. It's answering a riddle that is widely found on the internet – a fact the researchers admit. How is that common sense and not regurgitation? So they come up with their own new puzzles to test GPT-4, and it figures those out too. I have basic common sense (feel free to disagree), and I couldn't do most of these. The researchers assume that GPT-4 has a 'rich and coherent representation of

* The pictures really are terrible: the horn is all wrong, the legs don't connect to the body, and the face is too short – it looked magical, but in a bad way.

the world' because it knows the circumference of the planet is (24,901 miles), as though having that fact to hand and seeing how it slots into this riddle is a sign of AGI. Riddles are just word algorithms, not real life.

It's easy to laugh at this research, given the unicorns and bear hunts, but figuring out how AI understands our world is something academics have long considered. We don't really know when humans pick up commonsense knowledge, such as understanding that dropping an egg on hard ground means it'll smash into a mess. It'd be easier to tell if or when AI learns common sense if we had access to training data and models, but private companies aren't keen to share those.

LLMs have other problems in the here and now, if that end-of-the-world stuff doesn't do it for you. These massive models need to be trained on huge amounts of data. They've already pulled in the English-speaking internet, even though plenty of the words out there on the web are under copyright or are complete garbage, or both. To get more accurate, AI developers can tweak their systems, improving the model and the weights and how it all corrects itself, but the bigger the model, the more data is needed.

Where will it come from? Online publishers are pushing back: Reddit is asking to get paid to let LLMs mine its huge back catalogue of user-generated content; newspapers are considering lawsuits for copyright infringement; and governments are looking at how to make AI companies pay for such data, perhaps using the equivalent of the 'robots.txt' tool system that tells search engines a page shouldn't be crawled by their bots. It's easy to think of other banks of data: digital books, email and messaging accounts, and voice assistants, which transcribe what you say into written text in order to understand it. Those data sets could be sold to the highest bidder or used by the companies that already own them – feeling nervous about your Gmail or WhatsApp yet? – though that may be a decision for the courts. Expect lots of data rights lawyers to be readying for battle.

This potential data grab is a problem for privacy, but most digital technology innovations have privacy implications, be it your phone, computerised medical records or online shopping. But it also has serious implications for people who write for a living, and as that includes me, you'll understand the nervous tone of the next sentence. My publishers, current and former, own the copyright to my work, generally speaking. They will be more than happy to sell my out-of-date articles that no longer draw many readers to whom they can serve ads, but if I'm honest, they'll also cheerfully let AI developers hoover up my more recent work too. Why? Because it gives them another much-needed income stream. While I wish my employers nothing but financial success, GPT and the like are already being touted as replacements for writers like me – why pay a person (however poorly), when you can pay a company for use of their LLM to churn out untold paragraphs for $20 a month? By analysing my work, these systems are learning how to replace me.

Well, sort of. They're already capable of churning out dull marketing blog posts – 'top ten reasons why our product is just what you need!' – but real journalism, with its investigations, phone calls and going out in the world, can't be regurgitated. News Corp is already using AI to automate local weather report stories, but all that does is arrange words to say it's going to be hot and sunny today. Some people like a chart for that, other people want a sentence. It's hard to get het up about that. Humans, on the other hand, are still needed for the research side of journalism, but also for surprising, intriguing writing.*

* Mark O'Connell has my favourite example of this in his book about future technologies, *To Be a Machine*, where he randomly starts talking about pistachios because no AI would ever come up with something that silly. Although, once a neural network is trained on his book, that's no longer true.

If we use well-designed LLMs carefully to automate the boring, repetitive parts of journalism, that would free up reporters to go out in their communities, investigate wrongdoing and take time carefully crafting their copy. It's a cute fantasy, but as a journalist who's been made redundant for budgetary reasons more than once, it's always safe to assume that any money saved by technologies won't be invested in better journalism but used to prop up profits (or cut losses) instead.

But this is how AI could impact *me*. If you want real criticism of LLMs and the great AI panic of the early 2020s, there's one name you should know: Timnit Gebru.

When Hinton quit Google, he was awarded interviews in all the best publications, with glowing headlines dubbing him the 'godfather of AI'. But Gebru left Google years before, in 2020 – also over ethical concerns.*

Gebru and five co-authors were set to publish a paper that examined the downsides of large language models – it raised concerns about environmental impact and bias, nothing too controversial, really – but managers at Google asked her to remove her name and those of her colleagues, leaving Emily Bender, director of the University of Washington's Computational Linguistics Laboratory, as the only author. Gebru refused, and got told her job was gone, though Google's version of the story differs and suggests she left of her own accord. Jeffrey Dean, the head of Google AI, told employees in an email that the paper didn't 'meet our bar for publication'. Whether Gebru quit or was forced out, her departure was followed by that of her colleague,

---

* A short history of Gebru: born in 1983 in Addis Ababa, Ethiopia, she moved to the US as a teenager on political asylum. She studied at Stanford, where Fei-Fei Li was her PhD advisor, and worked at Apple, Stanford and Microsoft, where she co-authored a paper with Joy Buolamwini on bias against Black women in facial recognition software. Then in 2018 she was hired to work on AI ethics alongside Margaret Mitchell at Google. To be clear, this is an excellent CV.

Margaret Mitchell, co-leader of the Ethical Artificial Intelligence team, reportedly because she 'exfiltrated thousands of files'. Why do that? Again, reportedly because there was evidence of discrimination against Gebru, whom she was hoping to support.

Setting aside HR disputes, I'm going to point out something here that perhaps could be left unsaid: both of these individuals are, unlike Hinton, women. It's an intriguing pattern in AI that the companies and research are often led by men – white and Asian, predominantly – but that so many of the AI experts raising concerns and suggesting solutions are women, and often women of colour (it's worth also reading Joy Buolamwini and Ruha Benjamin on these subjects).

The paper written by Mitchell, Gebru, Bender and their colleague Angelina McMillan-Major was eventually published in 2021, though now only with four co-authors named. In the acknowledgements, the paper notes there were actually seven authors, but 'some were required by their employer to remove their names'. Margaret Mitchell was in fact not listed as one of the four, though a 'Shmargaret Shmitchell' was.

Their paper raises a few risks about large language models and their quickly growing size. Google's BERT language model in 2019 used a data set size that was 16GB; in 2020, GPT-3 required 570GB for training. All of that computing processing has a high financial cost, which limits who can build these systems, while the energy required has unquestionable environmental impacts, though the use of renewable sources is increasing in the sector. An average human's carbon emissions are about 5 tonnes (11,023lb) annually, but the authors say that training a transformer model (a type of neural network) using a technique known as neural architecture search (NAS) emitted 284 tonnes (626,113lb) of carbon emissions.

And the climate crisis is impacting those least likely to benefit from these AI systems. The paper notes that the

Maldives is expected to be underwater by 2100, and that 800,000 people were impacted by floods in the Sudan, asking if it's fair they pay the price of developing LLMs that aren't being developed in Dhivehi or Sudanese Arabic. It's past time for energy efficiency to be included in such models.

Another challenge the authors raised is data: it should go without saying that the internet is full of bias. As the paper points out, training LLMs that will have a wide and varying impact on a data set that remains largely written by men in English – Wikipedia editors are at most 15 per cent women, for example – will naturally overrepresent those ideas, styles of writing and so on. And then in turn those LLMs churn out more content that's identical, exacerbating the problem. Plus, as we rely on increasingly massive data sets, it becomes harder to document those data sets or even understand what they contain.

A third challenge concerns the very nature of LLMs: the paper dubs them 'stochastic parrots', meaning they mimic without understanding. The AI system is not applying meaning, that's up to readers to do. This is why so many people were surprised by 'hallucinations' or 'lies' in responses from ChatGPT – they didn't have the AI literacy to understand that such systems do not understand meaning or context, they merely spit out text that matches the pattern of our language. They do it incredibly well, and it fools us. These stochastic parrots create language without meaning. Any meaning in the output is supplied by the reader, and is therefore an illusion.

Bias, strengthening the status quo, misunderstanding the power of AI, environmental concerns and exclusionary costs – none of this is particularly controversial. But this was too much criticism for Google to handle – the same company whose CEO is doing the rounds warning about AI killing us all. What on earth is going on here? In response to that open letter from AI luminaries signed by Musk et al., the paper's authors (Gebru, Mitchell, Bender, and McMillan-Major) issued their own letter, saying: 'It is dangerous to distract ourselves with a fantasised AI-enabled utopia or apocalypse. Instead, we should focus on the very real and very present exploitative practices of the companies claiming to build them, who are rapidly centralising power and increasing social inequities.'

Surely the CEOs of AI-producing companies know all this too. So why are they signing open letters about how freaked out they are? There's a twofold theory, according to which the motivations are marketing (if AI is going to kill us all, it must be really impressive – so your company should sign up now) and regulatory capture (distract politicians by focusing on a way-off-in-the-future hypothetical threat rather than anything that will impact us now). It's a theory, anyway.

Now let's step back a bit. We've seen AI go through a seasonal cadence of summer and winter, hype followed by backlash, then quiet, real progress before the loop begins again. Is this the point where the hype crash happens again? Perhaps, but hopefully not. Instead, it'd be preferable for AI to break out of this cycle and start to be truly useful, with a careful, close eye on the downsides.

Let's ditch the historical hype cycle and build AI into a useful tool for people. Over the last several decades, we've refused to admit AI is narrow, when it is. Even before it could be built, we debated whether it could overtake human intelligence. And before we managed to put it to work, we shuddered over job losses. We should worry about existential threats and societal disruption, of course. But we need to fix the essential challenges facing us now, too – the costs (environmental and financial), the flaws we're building in by not taking our time developing data sets, and the wider misunderstanding of how these systems work and don't. People have been hyping AI abilities and threats for decades. Let's stop wasting energy on marketing and take the time and effort to make AI work well for people.

How do we do this? Better funding models, openness in corporate research and taking our time. But what we shouldn't do is listen to CEOs confounding regulators by screaming about existential risks as though the Terminator is standing right behind us. Lads, you're the ones making it. If you're so scared, stop what you're doing – and stop sidelining the people pointing out problems.

CHAPTER THREE

Robots

The woman waved her clipboard as the crowds grew around her. Flapping away on the exhibition floor at the ExCeL conference centre in east London, she sounded irritated: a key person was missing, no one was lining up in an orderly fashion and it was well past time to start. As it turns out, organising a robot parade is as difficult as herding cats.

This was ICRA 2023 – to the uninitiated, the Institute of Electrical and Electronics Engineers (IEEE) International Conference on Robotics and Automation – a major robotics conference featuring luminaries from Boston Dynamics founder Marc Raibert to cybernetics art critic Jasia Reichardt and body-double avatar maker Hiroshi Ishiguro, as well as startups, students, and inventors showing off their creations. Though everyone visiting ICRA had days to wander the booths looking at every robot set to take part in this parade, the appeal of seeing them stomp about en masse had gathered hundreds of attendees.

The organiser with the clipboard shouted the robot wranglers into position, found the missing luminary to grand marshal the parade and cleared the aisles for the procession. Just as the show was about to begin, a random man carrying a box stepped up to a robot and kicked it over before scurrying away, his motivation obscure; the robot was quickly set right.

The parade began to march. Many looked like smaller versions of Boston Dynamics' Big Dog – the four-legged one with backwards knees you've probably seen on YouTube – but those were joined by wheeled tanks waving jointed arms, tiny toy-like creations and a large, flat box like a mobile fax machine. A bat-like drone fluttered overhead for a few minutes. All of the humanoid robots used wheels rather than legs. The box-wielding robot kicker aside, there was no drama, nor any surprises during the parade. With a call of 'watch out, robots coming through', they trundled up one aisle of the show floor before returning down another, lined by people snapping photos and recording video (myself included).

Beyond bipedal omissions, there was something else to note: every parading robot was followed by at least one human, most holding what looked like a games console controller in their hands. And that's because none of these robots were smart enough to walk up and down a hall full of other robots and people without being directed like a radio-controlled toy car. That's the often-unrealised secret of most modern robots: we're afraid of them taking over the world, but most of them can't do anything beyond walking a few steps without our instructions.

Robots aren't the future – they're here and now and have been for decades, though what they're capable of (and not) may surprise you. Robots dominate production lines, assist in surgery, and vacuum our floors. What each of these examples has in common is specificity of use and a practical design to match, and that means their shapes don't resemble us humans. The robots building cars are giant jointed limbs that can lift and manipulate into place components both heavy and delicate. The da Vinci line of surgical robots wield multiple instruments in their precision arms, but are directly controlled by surgeons. And Roombas are small, flat suction machines, handily rideable by cats – and the closest most of us will come to a robo-butler.

Carefully developed for specific uses, these designs make sense. A humanoid robot pushing a hoover around is ridiculous and inefficient. Why not simply automate the vacuum? Yet the dream of an android, a humanoid robot, persists. We can't stop trying to build a do-it-all, mechanised person. And we have dreamed of it for centuries – Leonardo da Vinci himself in 1495 created a metal knight that could wave its arms and sit down to entertain at an Italian noble's parties. More than 500 years later, Elon Musk stepped onstage to announce his take on humanoid robots, though not with a working demo but with dancers dressed like robots; it looked like a student theatre performance. We just can't help it. To understand why we want robots that reflect our image, you'd need to study philosophy, religion and psychology – and you'd still be left scratching your meat-based head. Here, we'll just accept it as a persistent fact of humanity, and move on to how to make the darn things. Because that's a tough enough challenge.

Anyone who has watched an infant shift into their toddler years with their first wobbling steps knows that walking isn't easy. Unlike newborn deer or giraffes, up and running away from predators mere moments after birth, it takes us humans years to perfect this core movement. And it's not just walking: try not to scream 'that's not how you do it' when sitting with a toddler slamming the incorrect block into a shape sorter or opening the flap of a book by tearing it apart, and it's clear that precision doesn't come easy to fumbling humans. Learning more complicated movements – swinging a bat to hit an arcing ball, cutting a carrot without slicing fingers, pulling a bow across strings to make pleasing sounds – requires careful study and practice.

And that's despite our massive brains and well-evolved bodies. Imagine starting from scratch. The very first robots, if you can even call them that, arrived in the 1920s, with metal bodies and a few clunking motions, though some eventually included speech and one even smoked a cigarette. We quickly put them to work, or at least we did those robotic jointed arms that could manipulate components. These started working in 1961 in a General Motors (GM) car factory, and a few years later, artificial intelligence was added to a rolling robot in California so that it could navigate on its own – it probably says a lot about humans that we started with bodies rather than brains. In the 2000s, Boston Dynamics designed a robot soldier of sorts for the US military, and it all culminated in Japan with what was perhaps the first true humanoid android, Honda's Asimo.

Bipedal robots have done parkour (Boston Dynamics again), chatted to UN diplomats (Hanson Robotics' Sophia), and acted in theatre (Engineered Art's RoboThespian). That's on a good day – androids are as often laughable for their obvious flaws, with supercut videos of robotic gymnasts falling on their faces just as popular online as those where they land their backflips. But there's something lacking in robots: intelligence. We can programme robots to do anything, but we still have to tell them what to do – indeed, the latest models of humanoid robots are avatars that are remote-controlled from a distance by a human operator. And though sci-fi films spark dystopian nightmares and (as with AI in the previous chapter) fears of lost jobs, the reality of automated androids is closer to the Knightscope K5, a security robot that rolled into a fountain at a Washington, DC office in 2017 and drowned. Rest in peace, Steve the robo-guard. We'll never forget you.

★ ★ ★

A long 89 years before Steve's damp demise, William Richards had a problem. A very well-known public figure – often reported as being the Duke of York (later King George VI), though others say it was merely a renowned scientist – was booked to speak at the opening of the 1928 Exhibition of the Society of Model Engineers that Richards was organising in London, but cancelled at the last minute. Who could replace this lost-to-history luminary?

Richards had a solution: 'make a man of tin'. And that he did – well, he used aluminium. Eric, as the metal speechmaker was inexplicably named, wasn't exactly a robot as it couldn't walk but stood where its handlers left it. But its head and arms moved, and its eyes lit up – as did its mouth, with 35,000 volts of electricity sending blue sparks flying from its teeth to indicate anger. It's not clear exactly how Eric worked, but two people were required for its physical operation, suggesting it was more a man-handled, programmable automaton than a robot.

Where this metal man did show something approaching intelligence, or at least automation, was in speech. Eric could stand, wave and turn its head on verbal command,* but also answer basic questions, like a giant metal voice assistant. One journalist 'interviewing' the robot asked if it was married. 'But before he replied I knew it was a foolish question. He looked too happy,' writes the journalist, because this was the 1930s. Eric replied that it was 'not likely' to be married, 'because I am too young. I am only 18 months old.' Eric was 'so amiable' about all the questions that the journalist queried if it could ever lose its temper: 'He replied by staring at me with a fearsome look and spitting a shower of sparks at me.'

The voice magic was done through radio, but at the time Richards wouldn't reveal exactly how the question-and-answer sessions worked. 'All I can say,' he said, 'is that in constructing the "man" we have used the most advanced methods of radio control ... using some of the [Marconi company's] patents. I can assure you, however, that his speech is produced neither by phonograph record or talking film.'

Massive, limbed radio though Eric may have been, that was novel enough for 1928. Eric's initial four-minute address at the Exhibition of the Society of Model Engineers was a rousing success, so the metal man toured the UK, even deputising for the lord mayor in Newcastle, before heading stateside. The success of Eric and its contemporaneous robot rivals inspired others, including Elektro and its dog Sparko, which were created for the 1939 World's Fair, though again,

* You can watch Eric responding to verbal commands, and hear another out-of-date joke about why it's the perfect man, here: https://biturl.top/JFFBrm

purely as entertainment. But it was believed even then that human-shaped machines had the potential to take on jobs normally held by people: in 1933, a newspaper reported rumours that London County Council had ordered a humanoid robot from Eric's makers to photograph cars running traffic lights, though there's no evidence it ever took on that role. Westinghouse Electric Corporation's Televox, a similarly basic metal man that could respond to voice commands, stood in for watchmen at reservoirs in Washington, DC, perhaps marking the first jobs lost to robots, albeit just for the one shift. Given the robot could do little other than speak and stand, if a wrongdoer attempted to sneak past during the stunt, Televox couldn't have put up much resistance beyond a verbal ticking off.

But before Eric could get a job outside the world of entertainment, it disappeared – possibly for scrap – sometime after 1936, as that's when mentions in newspapers trail off. Several decades later, in 2016, the Science Museum in Kensington, London, rebuilt a version of Eric to star in its exhibition on robots. Because of Eric's disappearance, and Richards' reluctance to share its workings, Science Museum curator Ben Russell and roboticist Giles Walker had to make it up as they went, re-engineering Eric using modern components. To reproduce the voice in the modern Eric, Russell tells me he recorded snippets of his own speech on his phone for playback later.

Though Eric and its fellow metal friends were far from capable, reporters of the time saw potential. 'His actions may be limited at present, but so were the movements of the first cars – only 35 years ago,' noted the *Croydon Times* in 1936. 'If Eric can do all he can today, what will he do in 50 years time?'

A lot, as it turns out. While Eric was little more than an entertainment box built by a clever showman, by the mid-1980s, robots were starting to learn how to walk on their own – though it took some effort figuring out how to make legs work, and also to teach the machines how to think (or something like it). But as it happens, we had managed to put them to work well before then.*

★ ★ ★

Twenty years on from Eric, in 1956, the idea of putting robots to work was carefully discussed – not in a boardroom or a lab, but at a cocktail party in Westport, Connecticut. Entrepreneur Joe Engelberger happened to strike up a chat with inventor George Devol,† who was explaining his already-patented idea for a programmed article transfer device, a machine that could be taught how to handle objects in a repetitive motion. Impressed, Engelberger leapt upon the idea and convinced Condec to fund a joint venture with Pullman (the train company), which he dubbed Unimation (short for 'universal automation'). Based in Danbury, Connecticut, Unimation produced its very first commercial robot in just five years.

The Unimate One wasn't a walking, talking humanoid entertainer like its predecessors, but a machine built for work: it was a programmable, strong jointed arm that could pick up and manipulate items, specifically designed to take on 'strenuous, unpleasant or hazardous' jobs. The very first of these industrial robots was installed in a New Jersey GM plant in 1961 to pick parts out of a die-casting machine, meaning humans didn't need to go near the dangerously hot metal.

The Unimate was initially welcomed by workers, not least because it took on the worst jobs in car manufacturing – an idea that was pushed via marketing brochures with slogans

---

* The rebuilt Eric is in the collection at London's Science Museum, but check it's on display before visiting.

† Remember the New York World's Fair of 1939, where Bel Geddes showed off the future of driverless cars? George Devol was there too, showing off photoelectronics to automatically count customers instead of a turnstile. Devol was also one of the inventors of the Speedy Weeny, a microwave-style oven for cooking hot dogs.

like: 'shake hands with our new volunteer for the dirty work'. Unimate's advertising made sure to include humans in the loop, saying that 'it takes a man skilled in the details of a particular craft to direct and get the most out of this capable but still "blind" machine'.

GM paid about $18,000 at the time, though it cost Unimate $65,000 to produce that first robot; the company took nearly two decades from its founding to post a profit. Looking like a massive photocopier with a gun turret slapped on the top, like an office-bound tank, the 1.2-tonne (2,700lb) machine was 1.6m (5.2ft) tall and almost as wide, with an arm that mimicked that of a human, with a shoulder, elbow, and wrist that could lift parts up to 11kg (25lb) at regular speed or 34kg (75lb) at slower speed. The arm could reach out to pick up an object, and move it in an arc of 220 degrees around its body, 30 degrees upwards, and 27 degrees down, before setting it down again with 0.1cm (0.05in) accuracy.

Eventually, the 'hands' could be switched out to do different jobs: a sales brochure from 1962 explains that early models could switch from assembling parts to feeding a lathe, to welding, as well as loading a conveyor, operating a punch press, spray painting, or packaging products. To programme its actions, a user would hit 'record', then move the hands through the necessary motions. The robot would then repeat that sequence – a slogan from a 1964 brochure called for the user to 'just take it by the hand to teach it' – and it could store 200 such operations in a memory drum.*

The hydraulically powered arm eventually found its way into the factories of other American carmakers, and a more advanced version, dubbed the PUMA, made an appearance on Johnny Carson's *Tonight Show* in 1966, which often pulled

* This is so clever and advanced that a modern version of that style of physical programming was still in use 50 years later via Baxter robots – no programmers were needed, and the power of programming was in the hands of workers on the factory floor.

in millions of viewers. It poured a beer for the talk show host. (Slinging drinks remains an accomplishment used to prove the worth of robots, with mechanised bartenders serving at The Tipsy Robot in Las Vegas and on Royal Caribbean cruise ships.*)

Unimate should have been a resounding success. But after the PUMA poured drinks for Carson, show business remained the only role that corporate America could envision for the robot – Engelberger said the only calls he got after the Carson show were to hire the PUMA for county fairs. That was in the US, at least. Engelberger also had a call from a Japanese trade group that invited him out to speak to 500 industrial executives who wanted the robot for 'more practical applications'. They peppered him with questions over a five-hour-long session, and in 1969, Japan's Kawasaki Heavy Industries licensed the design to manufacture Unimate robots for that market.

Just like in the US, the Kawasaki-made Unimates first found their place among carmakers. Japan's automobile industry was booming – car exports exploded from 100,000 in 1965 to 1.8 million a decade later – but Japan lacked welders. That was one of the jobs the Unimate could tackle, with Kawasaki claiming the machine could do the work of 20 people if it ran day and night. Sales were held back by the

---

* While in California researching this book, my family ended up in a mall in south San Francisco (where we'd just been dropped off in the parking lot by Waymo's driverless car) and popped in to use the loos and grab a drink. In the food court was, by wild coincidence, a stand that served espresso made by a robot barista. Essentially a robot arm set amid coffee-making equipment, it could brew the coffee, steam the milk and combine it all. However, not only did it still require human support to reload supplies, but it also took significantly longer than a human barista, with a display on a tablet warning us it would be a several-minute wait despite no discernable queue. While it would have been faster to go to Starbucks, not for anything (even a fast frappuccino) would I give up the memory of my infant daughter yelling 'ROBOT BLEEP BLOOP' for the entirety of the several-minute wait.

first oil crisis and then the second. By 1977, Kawasaki had shipped just 500 Unimates. But then over the next three years, it doubled those figures to 1,000, and it was onwards and upwards. Eventually, Kawasaki did so well selling Unimates to local carmakers that it also set up shop in Detroit to widen its market, later expanding into Europe.

The robotic arm was an unquestionable success – but not for Unimation: its sales stalled, partly because the company insisted on sticking to hydraulic robots versus the newer electric models made by rivals. Even GM opted for the new designs, ending their partnership with Unimation after 20 years. Usurped by newer technologies, the company was sold off to Westinghouse in 1984 for $107 million, and is now in the hands of Swiss robotics company Stäubli. While Unimation has been subsumed, Engelberger and Devol's cocktail party-inspired idea changed industry: there are now 3 million industrial robots in use globally, many of which are specialist arms, just like their very first machine.*

★ ★ ★

Ten years after Engelberger and Devol teamed up, what constituted a robot had changed so remarkably that it's hard to believe how quickly it happened. Eric and the others like it were essentially metal puppets, their strings perhaps pulled via radio, electricity, or magnets, but not autonomous. The Unimate was programmable, but had no intelligence or ability to make decisions. That wasn't true for Shakey, a robot developed by Stanford Research Institute (SRI) from

* The first ever Unimate robot to be installed at that GM plant in Trenton, New Jersey is in the collection of the Henry Ford Museum of American Invention, though it's not on public display. The Smithsonian National Museum of American History also has a Unimate controller, from 1978, though it's also not on public view. It's a big box that takes up a lot of space, after all.

1966 to 1972: it could wander around a room, figure out where it was when lost, decide how to get around obstacles and even learn as it went. And modern versions of the systems Shakey used to manage all this have gone to the moon and are still used in cars today.

Unlike its predecessors, Shakey wasn't about the mechanics but the software. To keep it simple, Shakey didn't have an arm like the Unimate to manipulate objects but was essentially an electronics rack on wheels. Indeed, it earned its name because the shonky motors powering its rolling movement meant it shook constantly. But Shakey did have something special: artificial intelligence. The purpose of Shakey and its successors (including a follow-up with the equally undignified name of 'Flakey') was to build a testbed to integrate all the then-known subfields of AI – including problem solving, scene analysis, and natural language – into a single experimental platform.

That was what the researchers wanted to do, but when applying for funding from the US government, they also added another aspect to Shakey's job description: spy bot. Peter Hart worked on Shakey from the first day of the project until the day it ended in 1972, eventually leading the team. By his telling, the initial proposal to the Department of Defense's then-ARPA suggested Shakey would be 'a mobile automaton for reconnaissance' – they used the word 'automaton' rather than robot because, as Hart said in an interview for the Computer History Museuem's Oral History project. 'You have to remember that until Shakey, robots literally were fictional'.* But that was 'basically a cover story, and the real motivation was a testbed for integrating all of artificial intelligence technologies that then existed' – and a few that didn't yet, including computer vision.

* You can read the full interview with Peter Hart here: https://bit.ly/3RxQmWd and see CHM's Oral History collection here: https://biturl.top/aim2M3

The original project lead, Charlie Rosen, was an expert at extracting funds from military coffers for his work, regularly travelling to Washington, DC to shore up investment for himself and other researchers, including Douglas Engelbart, the original designer of the mouse. Rosen appears to have a sense of humour about his misrepresentations for cash: while he was sharing the idea of a mobile robot to test AI, a general asked if the automaton would be able to carry a gun. 'How many do you need? I think it should easily be able to handle two or three,' he reportedly replied – quite the promise about a shaking robot that didn't even have an arm.*

Physically, Shakey wasn't impressive. Now on display at the Computer History Museum in Mountain View, California, it comprises a box with cables dangling out at the rear, sat on a wheeled hexagonal platform that has a bar to push objects on the front and wires dubbed 'cat whiskers' along the edges to detect obstacles. A tower of hardware and cabling sits atop the box, housing a 120-pixel-by-120-pixel camera to 'see' alongside a laser range finder, similar to the modern lidar sensors used in driverless cars, and antennas to communicate.

'I remember when I first saw it I thought: "Gee that looks like a dishwasher on wheels",' said Helen Chan Wolf, who developed the image and location software, in an SRI video interview commemorating the project.† Shakey's final home at the museum has it spinning in a display case near much slicker-looking robots that came in the decades to follow. Though it looks basic and bashed together – because it was – Shakey was much smarter than the lot of them.

---

* Rosen's response came during a meeting with Pentagon generals to discuss the subject, according to John Markoff in his book *Machines of Loving Grace*. He later reveals that when Rosen was directly asked by a Pentagon official, Ruth Davis, about whether Shakey could replace a soldier, he admitted it wouldn't happen anytime soon but that he wanted to test the core ideas. Thankfully, Davis was also keen enough on the science to back the idea.

† You can watch that video here: https://youtu.be/7bsEN8mwUB8

Shakey could do things no other robot was yet capable of, including making decisions without human interaction. Here's how it worked. Shakey's world was an office space sliced into seven rooms using stub walls, which were tall enough that the robot couldn't just peer over them but short enough that humans could easily keep watch. Inside each room were ramps, boxes, and other objects to use in experiments. Shakey had a basic map of those rooms and where it thought various objects were, which it would update if a block or ramp moved – this was the model of the environment. Shakey also kept track of its own place on that map, mostly by monitoring its own motor movements; known as dead reckoning navigation, this is rather like counting your own steps.* However, it's a technique that allows errors to accumulate, meaning the robot would become less and less sure of its position. So Shakey had another location-finding tool: the camera would look at an immovable landmark in the room and see if that was where it was expected to be. To aid this, the researchers had the skirting boards painted dark to contrast with the walls. So when Shakey wasn't sure where it was, it looked at the corner of the room, made a line drawing of the skirting boards and compared that to what was in its own system. If it was out, it recalibrated. This meant that when Shakey felt lost, it stared at the corner for a bit.

---

* Shakey also used dead reckoning to poor use in another way. For a while, Hart recalls, every now and then Shakey would stop and pirouette doing a complete 360-spin before continuing on with whatever experiment it was completing. Programmer Mike Wilber eventually dug into the software to find out what was going on. For the first few years of its life, Shakey was connected not through a radio link but a cable, as regulatory approval for the wireless version took ages, and they wanted to get started. 'It turned out that buried somewhere in the code was a little piece of code that counted the revolutions that Shakey had made and would reverse it to unwind the cable,' Hart tells the CHM Oral History Project.

Shakey was also preprogrammed with low-level actions – rolling forward by a set number of feet and tilting the camera to look down, for example – as well as intermediate-level actions, such as going through a doorway or pushing a box. A bit of software known as the 'model of actions' acted as the physics for this seven-room world, holding the information about how those movements worked, in particular their requirements (to go through the doorway, you must first be next to the door) and their effects (go through a doorway and you'll be in another room). Those actions could be combined using a planning program called Stanford Research Institute Problem Solver (or STRIPS), which took the goal given to Shakey by the experimenter and worked out a plan for how Shakey could achieve it – and that was the real magic.

There's an example of what this all looks like in a video the SRI team made to publicise their research, which is on YouTube and well worth watching, not least for the jazzy soundtrack ('Take Five' by the Dave Brubeck Quartet)*. A long-haired Nils Nilsson, a Shakey project manager who would eventually be seen as a founding father of AI, taps commands into the teletext in basic but structured English, which is then translated into predicate calculus, the language understood by STRIPS. 'Use BOX1 to Block DOOR4 from ROOM2' would become a list of low-level and intermediate-level actions assembled by STRIPS:

GOTO (D5)
GOTHRU (D5)
GOTO (D3)
GOTHRU (D3)
BLOCK (D4, BOX1)

* You can watch the demo video on YouTube here: https://youtu.be/GmU7SimFkpU

Shakey already knew where it was but could also check for obstructions on its route using its camera, before slowly rolling around, pushing boxes to implement the plan given to it by STRIPS. But what if there were to be an unexpected obstruction? To exemplify this in the demonstration video, Rosen – unquestionably one of the engineering greats of AI* – runs grinning through the rooms dressed in a black flowing cape that looks borrowed from a Halloween Dracula costume, to dramatically knock a box into the robot's path. The gremlin, as the narrator calls Rosen, hasn't doomed the task; Shakey's cat whiskers feel the box in its way, its camera tilts down to examine and identify the obstacle. STRIPS then automatically, without experimenter interference, alters the original plan, and finds a different route or moves the box. When complete, Shakey sends a message back to the teletext: 'DONE'.

Shakey could also learn. A system called Macrops remembered the plans that worked and saved them for future use, letting STRIPS reuse aspects to solve similar problems, perhaps unblocking a doorway rather than blocking it. That saved as much as a third of the original time it took to compute the journey – no small thing when you're using very slow, room-sized computers.

If it all sounds simple now, that's because we're used to technologies formed around 'ShakeyTech', be it natural

---

* Hart describes Rosen as 'beyond phenomenal' and one of only a few people who deserves 'the overworked label of renaissance man'. Rosen wasn't the clichéd tech leader. He came from relative poverty in Montreal, waiting tables in the summers in the Catskills, where he met both his wife and an academic who advised him to try for an engineering scholarship, sparking the career that would lead to Shakey. But bringing AI to the world using military money wasn't enough to sate Rosen's curiosity about the world – he also started a successful winery and companies that delivered prescription drugs and let people make their own pickles. After retirement, he sat on the boards of several tech companies and the day before his death from pneumonia in 2002, reportedly recorded a tape of his thoughts on the future of AI, in the hopes of writing a book.

language processing as seen in Siri or Alexa, computer vision for driverless cars or navigation systems for your non-driverless car. The 'A*' algorithm developed by Rosen's team to help Shakey plan its route between stub-walled rooms is at the core of satnav, the path a video game character takes and even how Mars rovers get around.

By considering time, distance and other inputs, the A* model looks to minimise the sum of such costs, and does so with much less effort than other techniques, helped by the fact it only works a few steps ahead. Search for a route from Mountain View to Times Square, Hart says, and Google Maps has an entire map of the US to consider, while A* just works a few turns ahead; as soon as a potential route becomes longer than other options, that work is halted rather than waste time. It's clever and practical.

Given all that, Shakey is often cited as a precursor to drones and driverless cars, as no humanoid android or otherwise moving robot could exist without these ideas – even if Shakey did roll around on wheels rather than stumble on legs. 'Shakey was an autonomous robot that didn't do anything useful, like operate in a factory,' Rosen said at the time.* 'And it couldn't do anything very fast. But it taught a lot of people how to put together complex hardware and software systems involving computers and perception programmes, like vision, sensing, and interpretation, so that Shakey could operate in a complex way, by itself, without continuous monitoring by a human operator.'

---

* Courtesy of the Department of Special Collections, Stanford University Libraries. Rosen was interviewed in 1980 by Angela Fox Dunn, the famous celebrity profiler and niece of the founder of Twentieth Century Fox (now Twentieth Century Studios), William Fox, whose name lives on in the Fox News channel. This quote is from a pre-print copy of the text of the article that Fox Dunn sent to Rosen for corrections. He replied with: 'Kindly pardon the numerous corrections. Most of them are to moderate extreme statements. I hope you will take them seriously.'

Shakey earned plenty of media attention at the time. In 1970, a *Life* magazine profile called the boxy bot the 'first electronic person', erroneously claiming it could travel at 6.4km/h (4mph) – a mistake made possible because the journalist never even saw the robot in person. Annoyed, Rosen and team wrote a letter in response from the perspective of Shakey, with the robot complaining that the exaggerated account of its capabilities meant its existence was at risk because humans now feared it – Shakey begged for help from the magazine, saying it would even write a story for the magazine in exchange for a 'nickel's worth of electricity, a drop of light oil, and a cupful of distilled water.'*

So, Shakey set in motion plenty that informs robotics and AI even today. Perhaps it was picking low-hanging fruit, as Hart has humbly said, but the project was full of firsts in computer vision, navigation and more. Hart argues his colleague Helen Chan Wolf should be known as the Ada Lovelace of robotics, as she has a strong claim to being the world's first programmer of robots. Shakey also sparked careers that influenced the industry in other ways. Bill Gates tells a story about being shown the SRI video – the one with Rosen as a gremlin harassing Shakey – and driving south from Seattle to San Francisco to see how it all worked. Requests to see the SRI video came in from universities across the country, as well as from IBM, Unimation and GM, and researchers in Germany and Denmark. The video was considered so valuable that when Japanese researchers wrote to Rosen asking if they could view it, he needed to first ask permission of the US government before showing it outside the country.*

---

* A copy of the letter is held at the Department of Special Collections, Stanford University Libraries.

The Shakey project came to an end in 1972, as the government had little appetite for further robotics research and SRI lacked the funds to keep it going.*

After 21 years at SRI, Rosen went on to start his own company in 1978 called Machine Intelligence Corporation and, as a true renaissance man, he also co-founded a vineyard overlooking Silicon Valley. Unlike many people at the time – or indeed now – he didn't fear AI, robotics, or the combination of the two. 'The peculiar attributes of the human mind are not matched by artificial intelligence,' he said in an interview at the time.† 'I personally have zero fears about a machine replacing a full grown human – who is not senile.'

It took lying to the military and a genius dressed as a gremlin, but thanks to Rosen and his team, robots had started to get smart. Now they needed legs.

★ ★ ★

The bright yellow, four-legged robot trundles up to the closed door, arching up so its front-facing camera – let's not call it a face, okay? – can get a better view of what's blocking its way. Unable to open the door, the robot briefly turns towards the back of the room where another robot dog suddenly emerges, stomping towards the camera.

This second robot dog has a limb the first lacks: a multi-jointed arm fixed atop the front of its body, with a claw at the end. Straightening its backwards-facing knees for more height, the second robot stretches its claw up towards the door handle. Pausing, it inches its yellow body forwards

---

* Shakey was finished, but over at Stanford University at about the same time, John McCarthy added automation to a long-running project known as the Stanford Cart, starting a similar AI development platform for students to work with there.

† This is also from the previously mentioned Angela Fox Dunn story held in Stanford's archives.

before snatching the handle, turning it, and pulling the door open towards its not-a-face. This second robot dog sidles sideways out of the way, propping the door open with its round little feet before holding it fully ajar for the first robot to creep through.

This 2018 video toes the line of most Boston Dynamics demonstrations: impressive and intriguing, but also creepy and alarming.* Dexterity aside, it's particularly unsettling watching a robot help another, like witnessing the first step towards machines ganging up on us humans. One comment on YouTube notes: 'Somehow I think of the velociraptors in *Jurassic Park*.' And that's just for a robot opening a door for another; Boston Dynamics is famous for four-legged 'animals' and bipedal androids that can chuck themselves on to rooftops, leap through parkour moves and outrun us, sharing these capabilities through YouTube videos designed for virality.

When you think of humanoid robots, Boston Dynamics is likely what springs to mind, perhaps alongside a shiver down your spine. That's partially because of the viral videos, but also because the sometime DARPA-funded company sparked controversy and attention when it was bought (and quickly sold) by Google as part of the latter's X labs, amid debate over the tech giant's dealings with the US military, eventually swapping hands to Japan's Softbank and then Korea's Hyundai Motor Group. Boston Dynamics has since sworn off all military uses for its robots, saying they can't be weaponised.

Thirty years before that series of acquisitions, Boston Dynamics got its start at co-founder Marc Raibert's robotics group, known as the Leg Lab, which began at Carnegie Mellon University in 1980 before moving to MIT in 1986.

Raibert had studied neurophysiology at MIT, which may not sound like the right road for one of the world's

* You can watch the video on YouTube here: https://biturl.top/Jz2MNb

best-known robotics developers, but the course focused on the use of brain recordings and other sensor techniques to study the movement of animals and humans, including how they balance when walking. 'Some student who was a year older than me [advised] ... "you got to do what you like doing every day". And I wasn't liking what I was doing every day,' he said in his keynote speech at ICRA 2023. That meant that when he did find his calling, it was easy to make the switch. A visit to a professor's lab changed everything – he saw a robot arm in pieces, and it intrigued him so much he shifted his field of study. Raibert shows a picture from the era, of a young woman holding a robotic arm: 'In 1974, 49 years ago, I fell in love twice: once with Nancy, and once with robotics. Now, a few years ago, Nancy dumped me. But I'm still in love with robotics.'

After finishing grad school in 1977, Raibert had the chance to discuss his work with Ivan Sutherland, whom we'll meet in the next chapter as the creator of the first head-mounted display. Sutherland said he'd fund a robot. Which robot did Sutherland pick? A hopping machine.

That led to the Leg Lab, Raibert's own robotics project. The first creation was a planar one-leg hopper that could jump on the spot using a pneumatic system, bounce away at a set rate, avoid being knocked over when bumped and even leap over small obstacles, all using a simple control system. Subsequent robots added algorithmic controls, additional legs, and even an articulated spine to support running.

In 1991, Raibert's lab made its first animal-themed robot, mimicking not a dog but a kangaroo, though it looks more like a single leg hacked off a marsupial, rebuilt through spare parts kicking around a garage. The Uniroo had a single three-jointed leg with a steel coil spring at the ankle that stored energy to boost its bounce, and it could hop forward at a terrifying 1.8m/s for a minute – versus the average running speed of 5m/s for a human – demonstrating the skill in a short video on the Leg Lab website.[*] Even a decade pre-YouTube, Raibert and his team loved a demo video.

That same year, the team at the Leg Lab taught a bipedal robot to do a somersault, run, jump through a hoop and even pull a carriage. They later built bipedal robots that looked like turkeys (well, sort of) and could walk at 0.5m/s for a few steps, driven by an algorithm, as well as a flamingo-style robot that was built as a platform for experiments, so was designed to be super-reliable – it worked for three years before breaking a leg. 'Now in those days, we had to have a hydraulic pump in the room, connected by hoses [to the leg],' Raibert said in his keynote. 'We had a computer in the next room ... we didn't have everything on the robot.'

But, for Raibert, something was missing from his work, and it took the 1993 Wesley Snipes/Sean Connery film *Rising Sun* to realise it. The filmmakers borrowed two MIT robots, the Uniroo and the somersaulting biped, to feature in the film, and Raibert went with them from Boston to Hollywood. It took a month of steady work to prep the robots for a five-day shoot that resulted in a 15-second scene. 'It was so hard to take our robots out of the lab and get them to work,' Raibert says in an online interview. 'This was the first time we'd ever tried to take these robots that worked great in the lab ... out into the real world.'[†]

And by that, Raibert means turning the robots into products that people could actually use in the real world, without months of preparation.

In 1992, Raibert co-founded Boston Dynamics with Nancy Cornelius and Robert Playter. Initially the company developed simulation systems for the US Navy, but it went on to build a surgical simulator to train physicians. Though the system worked well, the company didn't believe it had the

---

[*] You can watch the very short video of the Uniroo here: https://biturl.top/zUZvaa

[†] You can watch the full interview here: https://biturl.top/ieuyIb

knowledge or marketing to make it a success, so killed it off. Not long after, Boston Dynamics pivoted to robotics, working secretly with Sony for years before creating BigDog.

The impetus and funding for BigDog came via DARPA in 2003 – which had also funded a portion of Raibert's academic work – with development starting in 2004. The aim was to build a robot that could carry a payload into the woods or desert, such as all the backpacks for a squad of soldiers.

That helped scratch the itch Raibert had felt in Hollywood – he and his colleagues had built a robot that could go out into the world and actually *do things*. Initially, the direction of research was influenced by customers, with Boston Dynamics trying to find clients that offered enough of an overlap with what they could do and wanted to do. So it's no surprise that all of the Boston Dynamics robots developed in this era feel a bit terrifyingly military. After BigDog came the RHex, a six-legged robot developed for covering rough terrain, which the tiny, flat bot accomplished by whipping its curved legs around on rotors. Then came Sandflea – my personal nightmare – a small, four-wheeled reconnaissance robot that could launch itself 9m (30ft) into the air to hop on to rooftops or, in my head, in through my window. There was also the less-charmingly named LS3, a BigDog-style robot for carrying soldiers' gear, with a round, barrel-chested body. And then there was Wildcat, a quadruped that could run 32km/h (20mph) on its creepily jointed legs, first warming up with a weird bouncing routine before galloping like a demented skeletal horse.*

Amidst all of this, DARPA also tasked Boston Dynamics with developing a humanoid machine that could be used to test protective gear and clothing – if you want to protect soldiers crawling through the dirt, you need a test mannequin that can do the same. That led to a $26 million project to

* You can watch this here: biturl.top/22ey6r. I still can't watch the video of Wildcat without feeling a bit sick.

make the Protection Ensemble Test Mannequin, or Petman for short. Video from the era shows Petman, with a flashing light for a head, walking on a treadmill, doing squats and – this being the military – pushups. Throughout the years, Boston Dynamics upgraded Petman to run at 7km/h (4mph) and gave it the ability to sweat – anything soldiers could do, Petman needed to mimic to accurately test equipment.

Raibert tells me at the ICRA conference that he never wanted to build humanoid robots, but the money was good. Indeed, after Petman, DARPA requested a humanoid robot to act as a platform for its 2012 Robotics Challenge; this eventually evolved into the smaller Atlas, the robot shown in the Boston Dynamics videos. Atlas remains what's known as a 'research platform' – a project that isn't being commercialised, but gets attention and funding because the work might spur other, more useful inventions. With hydraulic joints and 3D-printed components, Atlas can backflip its way through any situation by pairing its behaviour libraries (essentially preprogrammed skills) with sensors that see around it and models that understand how the robot will interact with the world. At the time, Raibert told the world it was going to be able to not just walk but talk, too. 'I'm like everyone else, I want to put on a show and get attention,' he tells me.

In 2013, Google acquired the company. As Google had no rush for a return on its investment - an idea referred to as 'patient capital' – that freed up researchers to do a bit more of what they wanted. That may well be why the number of goofy videos seems to have increased after that time.

Atlas was never commercialised, but its technology did evolve into a new product. Boston Dynamics wanted to build a robot to shift boxes, which it would then sell to the understaffed logistics industry. At first, the company simply programmed an Atlas to do it, but that wasn't fast enough; so researchers ditched the legs and turned Atlas into a rolling box-picker which balanced the lifting apparatus at the front with a battery at the back, looking like one of those drinking bird toys. Now Stretch, as it's known, is a big box on wheels with a single massive arm that ends in suckers designed to grab boxes and manoeuvre them into place.

Beyond Stretch, Boston Dynamics has another commercial product: the company has sold or leased out around 1,000 of the robot dogs named Spot,* the descendent of BigDog. Half the size, Spot was shrunk down on the advice of Google's Larry Page, who argued it'd be less scary to work alongside at a smaller scale. Spots can be programmed for a multitude of tasks, and are in use for research at universities as well as in dangerous real-life work roles such as monitoring oil rigs.

Spot is also used, more controversially, by police departments, largely within bomb disposal. But it isn't allowed to be used as a weapon. Despite its historical DARPA funding, Boston Dynamics signed an open letter alongside industry rivals pledging that their robots would not be weaponised.† The wording is loose enough to allow for BigDogs or their descendants to perhaps lug supplies for soldiers, but must leave the killing to humans.

Raibert still sees a future for humanoid robots, in particular in care homes or at-home care settings. But the videos released by Boston Dynamics don't show Atlas hefting seniors in and out of bed. Instead, it's doing the monster mash. There's a secret to Boston Dynamics' robots – well, it's not a secret, but it's easily forgotten: these robots are not acting autonomously but are very carefully programmed. Atlas can't watch a choreographer's steps and follow suit. It's a core thing to remember when watching any robot, dancing or otherwise. Raibert points to the phrase *if it walks like a duck, and sounds like a duck, then it is a duck*. 'Well, with robots it's not that way,'

---

* Raibert says Spot was named after the Eric Hill children's book, which has an innovation of its own: it was the first to have feature a lift-the-flap function.

† You can read the open letter here: https://bit.ly/3GwKJl9

he says, as robots don't have 'the intelligence, or the morality or immorality of a person'.

That doesn't mean there's no intelligence to these machines. The best way to understand how they work is to take a test drive, which I did with Spot at ICRA 2023. It's driven using a handheld controller with a tablet-style display in the middle. Tap the button, and Spot trots forward. It wouldn't know where to go unless you directed it, but it's intelligent enough to walk, even on rough terrain or up stairs; plus it can see and avoid obstacles – no matter how hard you try (and I tried), you can't make Spot trot into a wall. Raibert compares it to a radio-controlled car, noting that with such a toy you're in direct control but with Spot you make suggestions. 'And then Spot is trying to be benign, trying to go the direction roughly you say, balancing itself – you don't have to do any of that,' he says. When Spot runs out of power, or you're done with it, tap the correct button and it'll return to its charging dock, just like a Roomba.

Of course, if you know how to programme Spot, you can achieve much more than failing to walk it into a wall. During his keynote, Raibert introduced a pair of Spots to the stage to do the tango. While their ability to move is part of their systems, they had to be programmed to complete the dance, including holding a rose in a gripper hand. But the spectacle was cut short when the Spots fell over – though they earned a bit of applause when they managed to get back up on their own.

Such pageantry is marketing, but Raibert says he never expected to become a spokesperson for robotics, thinking he'd stay out of the limelight. That seems a surprise now, for a man who famously wears Hawaiian shirts every day, ensuring he'll be recognised after his keynote, where a queue of students, fans and customers wait to have a chat, share their own robots and snap a quick selfie. 'My plan of operation was … to just let the robots speak for themselves; nothing mattered about me,' he tells me. But he's now a figurehead for AI as well as for robotics, having moved into Boston Dynamics' AI Institute.

There, his work is to make robots smarter: not just in how they move, such as helping Atlas perfect its backflip, but also enabling a robot to learn by watching us, perhaps asking questions to fill in the gaps. In other words, Raibert's not quite done with legs, but it's time to focus on the brain now.

★ ★ ★

Katsuyoshi Tagami spent 10 years working on a secretive robotics project at Honda – and it made him anxious. His bosses at the famed motor company wanted a project to inspire their engineers, hitting on a humanoid robot – first, because they're beloved in Japanese culture, and second because the company believed that it would be able to navigate homes better than wheeled models. In 1986, Tagami started by looking around at what everyone else was building. He told the *Wall Street Journal* that he visited a lab – clearly Raibert's – full of six-legged and four-legged robots, even a one-legged hopping model – but no two-legged versions, and that worried him.*

It was soon clear that Tagami needn't have worried – technically, at least. His team of engineers at Honda began with the E-series, for 'experimental', which looked like a small rectangular appliance, perhaps a toaster or microwave, sitting atop a pair of robotic legs. Powered by connected hydraulic cables, the E0 could walk in a straight line, but only at a slow pace, taking five seconds for each step – try doing that in your own hallway and see how long it takes before you fall over or get bored. Over the next five years, Honda developed three more models – the E1, E2 and E3 – which managed to up their pace to 3km/h (1.8mph). Better balance and posture controls were added to the next three robots

* You can read the story on the *Wall Street Journal* here: https://biturl.top/qEb2eu

– E4, E5, E6 – which allowed faster walking speeds as well as the ability to walk up stairs or step over objects.

Then, a design change. No longer was the robot just a legged box, as the P-series – for 'prototype' – added an upper body and arms, creating a human shape that could walk on two legs. After just a decade of work, in 1996, Honda had a bipedal robot. But that success sparked another concern for Tagami and Honda's then president Hiroyuki Yoshino: was Honda playing God – and more concerningly, would others, in particular Americans, think they were?

Japan loves robots. Some pin this on the dominant Shinto religion, as it includes the idea of animism, the belief that all objects have a living soul. Others suggest it's merely down to the success of the cultural phenomenon *Astro Boy*, which began as a manga series about a robot fitting into the human world. If this sounds like a bit of a weak argument, perhaps it is, but if we consider the cultural artefacts around humanoid forms in the Western world, either as inspiration or reflection, they're not very positive in comparison. Mary Shelley's *Frankenstein*, the *Terminator* series and Czech writer Karel Čapek's 1920 science-fiction play *R.U.R.* (*Rossumovi Univerzální Roboti* or in English, Rossum's Universal Robots), in which the robots revolt against us humans.

Either way, Honda started the project to help motivate its team of engineers – simply having a robot in development gave meaning to the company's other work. But Honda spotted that Americans were worried about job losses and, you know, dystopian outcomes (hence the mass popularity of films like *The Terminator*). With such concerns in mind, Tagami went to Vatican City. There, Tagami met with Rev. Joseph Pittau, an acquaintance of company founder Soichiro Honda, for religious advice.

They walked, the story goes, to the Sistine Chapel, where Pittau pointed to Michelangelo's depiction of God creating man – the famous, ceiling-sited fresco of the bearded, robed God touching fingers with a very muscular Adam. The reverend told Tagami that building a robot wasn't playing God, but 'using your imagination to make something useful', as the *Wall Street Journal* reports – though it adds that Pittau later clarified he was unaware that the robot took the shape of a human.

That was blessing enough, and Honda's work continued. One goal was to reduce the size and weight of the robots: the P1 was a whopping 1.9m (6ft 2in) tall, but by the time the P3 rolled around it had been reduced to 1.6m (5ft 2in). A year after Tagami retired in 1999 came Asimo. Much smaller than its predecessors at a childlike 1.2m (4ft), it was built short enough to be useful but feel unintimidating.

Finally, a commercial, humanoid robot that could walk, talk and even choose its route as it did so, this robot helper was the first to be shaped like a human and think for itself, if only a bit. All of the Honda robots before Asimo walked using stored patterns, but Asimo decided on a pattern as it went, which meant it could change its steps or turn in a more natural way, shifting its centre of gravity as it went, in a way more similar to humans. By 2005 it could carry a tray and even hold hands with a human, as well as run at 6km/h (3.7mph). Over the years, Asimo learned how to return a kicked ball, shake hands and distinguish between three different people speaking to it at the same time. It eventually topped out at a 9km/h (5.6mph) running speed, about average for a reasonably athletic adult.

Asimo quickly became a star, but – like Eric several decades before – not for doing any real work. The robot rang the opening bell at the New York Stock Exchange, conducted an orchestra, appeared on British trivia show *QI* carrying coffee out to the presenter Stephen Fry, and walked the red carpet alongside actress Amanda Bynes for the Dreamworks animated film *Robots*. But though it could walk and talk, its capabilities remained seriously limited. Entertainment seemed to be its only purpose and Honda refocused its robotics on smaller, specific-use helper robots, including a mobility aid that can carry people and a companion robot shaped like a bowling pin with an expressive face. Humanoid robotics weren't over for Honda, but the company instead switched to remote-controlled avatars. Like its originator Tagami, Asimo was due for retirement: in 2018, Honda halted development. So long, Asimo.

Honda wasn't the only Japanese company unveiling walking humanoid robots – and there was a reason for that. The Japanese government in 1998 launched a project calling for industry to develop humanoid helper robots for two reasons: to spark innovation and to find a technical solution to a lack of carers for Japan's rapidly ageing population. Alongside research at various Japanese universities, industrial and technology giants also stepped in.

Between 2001 and 2005, Fujitsu had a mini humanoid robot series whose models were below 60cm (2ft) tall, while Hitachi unveiled a rolling bot called EMIEW (Excellent Mobility and Interactive Existence as Workmate) that could talk and easily navigate smaller spaces using a clever swinging body. Mitsubishi's cheerfully yellow rolling robot Wakamaru was designed for home use, but the high price of $15,000 and limited capabilities meant few were sold. Eventually it was rented out as a receptionist for the equivalent of $1,000 a day – more than you'd pay for a human to do the work.

In 2000, Sony followed suit with a tiny bipedal robot prototype called SDR-3X that was just 50cm (1ft 8in) tall, though perhaps better known was the company's AIBO robotic dog, which was released the year before. The AIBO was a successful, and adorable, commercial product that sold more than 150,000 units and eventually received upgrades including speech and image recognition. The AIBO was discontinued in 2006 but was eventually revived in 2017 – which, alongside Boston Dynamics' Spot, suggests there's something to be said for designing robots that don't look like humans but like our favoured pet.

In 2004, and after two years of development, Toyota unveiled a 1.2m (4ft)-tall prototype robot, amid additional efforts to build machine assistants for elderly people or to take over heavy labour. Though it showed off more practical skills, like walking, it also carried a trumpet and played 'When You Wish Upon a Star' during its demo – a way to show off its super-flexible artificial lips. By 2007, then president Katsuaki Watanabe believed that within a few years such 'partner robots' would be ready for industry and even homes; indeed, a rolling version called Robina with a head mimicking a bob haircut was already in use on Toyota sales floors to help shift Prius cars.

In 2017, Toyota unveiled the third version, dubbed the T-HR3 (for 'Toyota Humanoid Robot'). However, it was controlled not by computer intelligence but by a remote operator, who wore a head-mounted display while sitting in a 'Master Manoeuvring System' with arms, shoulders and even legs enclosed to capture movement.

* * *

While all of this impressive work was happening in Japan, a French company was building its own humanoid robots. Bruno Maisonnier's Aldebaran Robotics is the brains behind two of the most successful (or at least best-known) robots in history: Nao and Pepper. Launched in 2008, the dancing, football-kicking, research-enabling Nao is one of the most successful bipedal robots we've yet to see.

While only a wee thing at 55cm (22in) tall, it was enough to nab the attention of billionaire SoftBank owner Masayoshi Son. Sometimes called 'the Japanese Bill Gates', Son studied in California before returning to Japan and launching SoftBank, which began life as a telecoms company before evolving into a tech investor. Son wanted to push forward his dream of humanoid robots, and SoftBank bought a majority stake in Maisonnier's firm in 2013.

But money comes with requirements: after the acquisition, development on Nao had to take a backseat to Pepper,* as Son wanted a prototype in just three months. Rushing a technology that no one else has been able to make work is never a good idea. To make it possible, the company reused some of Nao's technologies on a larger scale. And it worked – sort of. For Aldebaran's first demo of Pepper in Tokyo, the prototype did a little dance, sparking a smile 'like a kid' from Son. But it was another two years before Son could show off Pepper to the world.

Pepper differs in a few ways from Nao: it's taller at 1.2m (4ft), lacks legs in favour of wheels and features a chest-mounted tablet to aid human–robot interactions. It was designed to be a 'robot with a heart', with emotional understanding. Like Nao, Pepper is actually in use: it has found work as a hotel receptionist, a COVID-era cheerleader and an assistant helping weary travellers find their way in airports, aided by its 15 languages. But though it only cost $1,790 to buy the robot itself, a monthly subscription of $360 made it a tough sell, for homes at least.

And Pepper had limited skills – it couldn't unload a dishwasher, let alone clean anything else, so it's clear it wasn't much use as a support robot – and even the tools it did boast often failed. According to a report in the *Wall Street Journal*, one funeral home trialled Pepper with reading Buddhist scriptures at ceremonies, but it repeatedly failed test runs.† A nursing home used Pepper to lead exercises for its residents, but limited moves and mechanical errors cost it the job. One journalist bought the robot for in-home use, but said Pepper's much vaunted facial recognition simply didn't work, concluding after a three-year trial that the robot was 'a waste of money'.

---

* A subsequent bipedal robot called Romeo, at 1.4m (4ft 5in) tall, was shown off by Aldebaran as a research platform, but it's since disappeared from the company's website in favour of Plato, a more practical, rolling bot, designed to help service workers with lifting and carrying.

† You can read the *Wall Street Journal* story here: https://biturl.top/AjqAVn

Pepper was one of the most advanced robots yet to hit the market for general use, and it was clearly not much use at all. If it can't remember who you are, read out pre-programmed text in a replicable manner or offer any actual support around the house, it's a technical marvel without any real purpose – just like Eric before it. Unsurprisingly, Pepper never evolved into the success either Son or Maisonnier dreamed about, and in 2021 production had halted. The end result was SoftBank selling Aldebaran to Germany's United Robotics Group in 2022, which is continuing to sell both Pepper and Nao.

★ ★ ★

Another line of research, away from legs and brains, has been faces. Hanson Robotics' Sophia has lifelike-ish skin and features, on its face anyway, making it eerily human – until it speaks. When Sophia was brought on stage at ICRA, the voice was stilted and, well, robotic.

Engineered Arts in Cornwall, UK makes robots to replace actors, with 150 of its RoboThespians in use since 2004. Instead of the weirdly human skin of Sophia, facial imagery can be projected on to the inside of its otherwise bright white face to change how it looks. The eyes can be animated, and limbs and hands offer a wide range of expressive movements. Controlled either by remote operator or pre-programmed, the robot has performed in plays, including one in New York called *Uncanny Valley*, alongside a human actor.

But perhaps most beguiling is the work on human-like looks done by Hiroshi Ishiguro of Osaka University in Japan. One section of his ICRA talk was given by his 'Geminoid' body double, a robot that from the distance at which I sat was almost indistinguishable from him – complete with his face, thick floppy hair and head-to-toe black ensemble.

The delivery of the speech was a bit stilted, but if I'm perfectly honest was better than a nervous public speaker such as myself could manage. But like Eric so many decades before, the robot wasn't actually speaking – as in forming the words itself – but instead playing back a recording made by Ishiguro, while moving its lips and gesturing its hands at an appropriate time.

★ ★ ★

The black-and-white-fabric-clad robot slowly marches onstage, three lights demarcating its machine face – before suddenly dropping the charade and breaking out into an enthusiastic, and very human, dance. Despite this being the Tesla bot unveiling in 2021, it isn't a robot at all, but a human dancer in a not particularly convincing robot suit. There is nothing funnier than a bad demo, and Elon Musk is the master of them, having once shattered on stage the window of a Tesla truck that was supposed to be bulletproof.

Subsequent demos were much better: an actual prototype of the bot, now called Optimus, was unveiled a year later, taking a few steps and doing a little dance. Musk promised that eventually Optimus would be able to have a chat, do household chores and work in a Tesla car factory. At a physical level, he said Optimus would be slow enough for you to run away from and weak enough that you could 'most likely overpower it'* – once again invoking the fear of the robots that we can't help but try to build. The aim is to eventually deliver the robot at a price below $20,000, borrowing from Tesla's autonomous car and AI work.

Whether Musk delivers on that promise or not, he has plenty of company. The Tesla bot's demo seems to have sparked a fresh rush to market, though of course Asimo's retirement didn't stop robotics research at Honda or anywhere else. Indeed, Sony has said its R&D department has advanced the technology to the point that it could start churning out commercial humanoid robots, but just doesn't see the point as these robots have no purpose. 'In terms of technology, several companies in the world, including this one, have enough

* You can watch the demo here: https://youtu.be/HUP6Z5voiS8

technology accumulated to make them swiftly once it becomes clear which usage is promising,' Sony Chief Technology Officer Hiroaki Kitano told reporters. 'The key is the development of application.'

That history of humanoid robots failing to find a purpose, let alone the difficulty of building the technology, hasn't dissuaded a swathe of new companies – nor their investors. As this book is being written, a host of startups are toiling to overcome that legacy of failure. Sunnyvale, California-based Figure says its AI-driven robot, Figure 01, is designed to fill in for the lack of warehouse workers; a prototype has managed to take a few steps, but also gave the middle finger to visitors using its dexterous, multi-jointed hands. Norwegian-based 1X Technologies has a rolling robot called EVE that has some AI functions, with the rest managed by a remote human operator; the company is also working on a bipedal robot called NEO with GPT-maker OpenAI. Meanwhile, Canada's Sanctuary AI has Phoenix, which can walk and carry heavy loads, all controlled by AI, though it's not a general-purpose machine and works in specific settings. Agility Robotic's Digit is bipedal but not particularly human-looking, with bright LED eyes set in an otherwise faceless tiny head, flipper-like grippers rather than fingers and backwards-facing knees. Designed as a co-bot to work alongside human employees, this 'multi-purpose' machine can so far only do one thing: pick up and move totes, the plastic boxes that are widely used in warehouses – though one Digit fell over and collapsed multiple times while lifting totes at a demonstration at a supply chain exhibition in Chicago. And then there's Shenzhen, China-based UBTech. Its 1.2m (4ft)-tall Walker robot can respond to voice commands, walk up the stairs and do yoga, as well as household chores such as serving tea, hoovering and wiping the counter. Ask it to bring you a beer and it'll stomp over to the fridge, open the door, bring back the bottle and even open it for you. The Walker has been put to use in exhibitions, where it sported a panda head.

What these companies, along with numerous rivals I haven't detailed, have in common with their predecessors throughout the decades is as yet they largely remain in development and haven't delivered on their many promises. That doesn't mean they can't or won't, but when it comes to humanoid robots, making ones that work in the real world is no easy feat. Right now, the main role for bipedal, human-ish robots seems to be serving as research platforms, with entertainment or marketing gigs on the side, and the odd trial here and there in retail, security and logistics.

That's a smart move, as the logistics industry is understaffed, as it's not often well paid and is usually hard work. However, it's worth noting that many warehouse robots are already in use without looking anything like humans. Britain's Ocado Technologies uses thousands of rolling box robots to shuffle grocery orders. And though Boston Dynamics began its work developing a warehouse robot with the bipedal Atlas, it slowly evolved from the humanoid form to Stretch, with its rolling design and AI to understand what boxes to pick and where to put them.

When you're reading about humanoid robotics developments, there's one question to ask: autonomous or avatar? As is so often the case with apparently intelligent machines, there's a lot of human graft behind some of these robots, and many are remotely controlled by human operators. That can make sense: why send a human into a melting-down nuclear plant, deep mine or offshore oil rig when you can have them control a robot from a safe distance, getting the best of both worlds by combining a human mind for decision making with a robotic body to avoid risk of injury?

For example, after retiring Asimo, Honda shifted its robotics development to what it calls the Honda Avatar Robot, a humanoid machine that's remotely controlled by a person. While there's still some intelligence involved – it needs to learn to walk and see, for example – decisions about what actions to take are left to the human operator.

Remote control could also be how we finally get robot butlers. UK-based Prosper Robotics, headed up by former OpenAI staffer Shariq Hashme, promises its Alfie wheeled robot can cook you dinner and clean the dishes – but there's a catch. Alfie isn't an autonomous miracle, but an avatar driven by a remote operator wearing a VR headset in the Philippines. Maybe it's just me, but I would rather do my own dishes or hire a local cleaner than allow a robotic remote worker into my kitchen, where the knives and oven are kept, but to each their own.

So far, autonomous robots aren't very good in the workforce. Boston Dynamics' Spot has found a niche in dangerous situations, making its cost and the high effort to programme for deployment worth the trouble. But you wouldn't buy one for restocking socks in a shop, unless it was a marketing ploy. Just as Pepper struggled to hold down a job, so too have other robots that have taken human roles.

In Japan, the Henn na Hotel nabbed a spot in the *Guinness Book of World Records* as the world's first robot hotel, plus plenty of media coverage too – 'Robot Raptor Receptionists' is one heck of a headline, after all. But after just a few years, the odd little tourist trap fired half of its 240-plus-strong fleet of machine helpers, because they made more work than they did, meaning the hotel had to ask human staff to work overtime, according to a report in the *Wall Street Journal*.* The 'Churi' room-based robot assistants would wake up customers in the middle of the night with 'Sorry, I couldn't catch that, could you repeat your request', their voice assistant mode awakened by snoring. The puppy-bot dancing welcoming committee was removed from the lobby after breaking, and the robotic bellhops couldn't actually reach every room, as they only worked on flat surfaces and would get stuck next to each other in the hallways. While some of the robots did appear to be useful, it wasn't the humanoid

* You can read the WSJ story here: https://biturl.top/iMvaAf

models. Indeed, one bipedal robot did nothing other than play piano in the lobby – why not just get a player piano, they've been around since 1896, after all.

As mentioned earlier, one place Raibert and others predict humanoid robots will have an impact is in care homes, or giving care to elderly or ill people in their own homes. As with logistics and warehousing now, and welding back in 1970s Japan, it's an industry that's short-staffed. But unlike those jobs, such robots would be on the front line of service, interacting directly with people, sometimes in their own homes.

As with the other two industries, it's clear to see how robots could assist staff in care homes: delivery bots could shuffle meals and other items around, saving steps, energy and time, while robots could be used to help support lifting and moving patients. One council in the UK has trialled the use of exoskeletons to make it easier for a single support worker to move a patient, while plenty of companies have non-humanoid robots or machines to carefully lift bedridden people up and into a wheelchair; and Japanese developers have created the Robear to carry people in its arms – forget humanoid, let's just make cute bears. But wholly replacing a human care worker with a robot, rather than a human carer helped by exoskeleton, leaves elderly or ill people entirely in the care of a machine, which is not only technically harder but sits less easily on the conscience. Would you feel guilty leaving grandma with no company but an Asimo most of the time? Perhaps it's better than nothing, if you're not planning on visiting anyway.

The rush of announcements and startups and demos means it feels like an exciting time for humanoid robotics. But don't forget it's been a long, slow road to get to the point where robots have the intelligence to take steps – and backflips – using their own skills. When an actual robot is on display, consider whether it's an expensive prototype or production-ready build, as plenty of robots get built and demonstrated but never manufactured. And when you see a clever video, remember they are theatre, as full dance routines take months of programming or, in the case of the Tesla bot, a human stand-in.

It's taken almost a century to get from clunky Eric to partially automated robots rolling around a London conference hall. And though there's work to be done before humanoid robots with some degree of general intelligence can join that parade, they will exist one day – simply because we'll clearly never stop trying to build them.

CHAPTER FOUR

Augmented Reality

It was one of the most dramatic product launches in tech history – or at least the early 2000s. Sergey Brin walked on to the stage at San Francisco's Moscone Centre at the end of June 2012, looking as much like a character from *Mission: Impossible* as is possible for a Silicon Valley CEO. Dressed head-to-toe in black, the Google co-founder sported on his face a black pair of smart spectacles known as Glass, the technology he was about to unveil at the company's annual developer conference.

'This can go wrong in about 500 different ways,' said Brin. How right he was – not for the demo, but for Glass itself. Rather than the usual video demo, Brin casually mentions he's loaned a Google Glass kit to his friend, 'JT'. 'He's actually pretty close by – he's just about a mile overhead right now,' Brin says. Brin joins a video call with JT, asking if he could get that loaner unit back, as though this isn't an overplanned stunt watched by thousands.*

Brin's dear friend JT looks out of his aircraft window, projecting to the auditorium a marvellous, though blurry, live view of the city by the bay from the glasses on his head. The crowd cheers; and then JT drops out of the plane, streaming the whole skydive down before thumping safely onto the roof of the Moscone Centre – the very building where Brin is delivering this demo. Another extreme sports dude takes the Glass kit, rappelling down the side of the building to the third floor, where a mountain biker grabs the hardware and jumps up the stairs and onto the stage. While

* You can watch the full reveal on YouTube here: https://youtu.be/D7TB8b2t3QE

streaming sports videos may not sound all that exciting now – or even then, as JT's flight suit bore the logo for sports-camera maker GoPro, though that action cam didn't get live-streaming capabilities for six more years. But it was still one heck of a stunt at a conference dominated by the latest features in the Android development kit and Google Docs' new offline editing mode.

At the time, I was the news editor for a British computing magazine, *PC Pro*, though I had to wait two years before Glass landed in the UK and I finally got to use it. Pulling off my non-smart glasses, I slipped the orange-red Glass glasses on to my face. (We struggled to know what to call them: Google Glass glasses sounds ridiculous.) Glass had two controls: voice and tapping the side button. But no matter what I did, all I could see was a blurry image off in the distance of my viewing range, like watching a broken TV from across a large living room. Or, as one review said at the time, 'like seeing a 5-inch smartphone hovering horizontally a few feet in front of your face'.

Our £1,000 investment wasn't broken: that was how Glass worked, when it worked – and there were bugs. The voice tools couldn't understand some British accents, it ran hot when Bluetooth was enabled and the touchpad that ignored my fingers could be activated by floppy hair – Brin wasn't wrong when he said Glass could go wrong in 500 ways. A year later, Google pulled the plug, and Glass was nothing but a punchline.

Augmented reality (AR) is where digital information is overlaid on reality, so you can still see the world around you. But these headsets have struggled to sell due to cost, battery life and expectations. Forget the 'hologram' effect of digital content floating in the world, AR looks like a broken, hovering TV. Virtual reality (VR), meanwhile, immerses you in a virtual world, such as a game, where you can't see anything outside the headset. These headsets still cause motion sickness for some, but costs have come down and VR headsets are now available at any tech store for less than the cost of a smartphone. The first time I used Glass, I panicked that I broke it. The first time I used a prototype Oculus Rift, I felt ill.* Such woes are why AR smart glasses and VR headsets remain niche toys rather than mainstream devices, and why the user experience fails to live up to expectations – though disappointment is inevitably a risk when the hype machine begins with a guy jumping out of a plane.

VR is unquestionably here, but AR continues to be a challenge to develop. And no wonder, as layering digital data onto reality in a lightweight form factor with all-day battery life is more complicated than producing a console-powered headset for playing games alone in your basement. And we've barely made VR headsets that work despite development stretching back to the 1960s, when Ivan Sutherland and his team of students built the first head-mounted display.

★ ★ ★

The first person to use what would eventually become known as an augmented reality headset didn't want the darn thing on his head. And it's no wonder he was concerned. As a student at Harvard in 1968, Quintin Foster was working under Ivan Sutherland, an academic already famous for his *Sketchpad* program that let the user draw as an input to a computer, a breakthrough that eventually led to graphical user interfaces, object-oriented programming and the field of human–computer interaction. Not bad for a PhD thesis, which not only earned Sutherland his doctorate but also the 2012 Kyoto Prize and the 1988 Turing Award.

Sutherland's next invention was a head-mounted display which used twin 25mm (1in) CRT monitors to bounce

* My friend Dean, who was into gaming and had a 3D world-building startup, got his hands on a Rift in the very early days. To give Oculus its due, Dean thought it was super-cool and continued playing while I sat on the sofa, clearly where I belong.

computer-rendered 3D images into the wearer's line of sight. This required 3,000 volts – and shocks above 2,700 volts are usually enough to kill or at least severely maim. 'I remember very distinctly having a discussion with the technician – "who's going to put this thing on?" – and I lost the argument, so I ended up wearing it,' Foster said in a panel discussion to mark the 50th anniversary of the project.* His face is pictured, obscured by the headset, in the now-famous papers on the head-mounted display released by Sutherland and his collaborators. 'I wasn't scared,' he says, 'I was apprehensive.'

Unlike *Sketchpad*, Sutherland didn't give this creation a snazzy name. Indeed, he makes it very clear that the device should be known as a 'head-mounted display' (HMD), and not the 'Sword of Damocles' – though that's how many misremember the name, it's just one part of the apparatus.†

Sutherland came to this work through a dislike of his first field of study: engineering drawing, which he studied at Carnegie Mellon. 'It seemed to me that having to erase dirty marks on paper and leave all that eraser crud was a bad thing, and if you could just have a nice clean computer do it, the lines would come out straight, and it'd make a nice drawing,' Sutherland explains in the panel discussion. Hence, *Sketchpad*.

Afterwards, Sutherland worked for the National Security Agency (NSA), the Pentagon, and ARPA as part of the Reserve Officers' Training Corps (ROTC), through which the US military pays your college tuition in return for two years of service post-graduation. He then moved on from the military to work at Harvard University as an associate professor of electrical engineering, where he tasked a handful of students – including Danny Cohen, Quintin Foster, Chuck Seitz, and

* You can watch the full panel discussion on YouTube here: https://biturl.top/FRvY7f

† Sutherland doesn't do press interviews, but still took the time to reply to my email clarifying the name and asking me not to screw it up.

Bob Sproull – to help build a head-mounted display that reacted to where the user was looking.*

Sutherland had come up with the idea for the project years before. Indeed, in his famous paper called 'The Ultimate Display', published three years before the summer project, he detailed what he thought this technology, not yet even named virtual reality, would be like. The user could move around, with the position of their eyes tracked, impacting the scene in real-time. They could manipulate objects and move according to real physics, and, he suggested, one day perhaps they could even die if shot in a game, once technology had come on some. 'Don't think of that thing as a screen, think of it as a window; through that window, one looked into a virtual world,' wrote Sutherland.

Sutherland claims his main contribution to the project was to realise that they didn't need to redesign the wheel. Thanks to his military connections, he was aware of a head-mounted display designed by Perkin Elmer for Bell Helicopter, which allowed the pilot to see directly below to help with landings, using a camera at the nose of the helicopter. The pilot couldn't remove their helmet and communications gear to see the feed, so it needed to be shown in front of their eyes. When the pilot moved their head, the camera turned too. That was a simple camera feed, though; Sutherland wanted to show a virtual, computer-generated image, rather than reality.

Handily, Bell Labs had a standing annual donation of $25,000 to the dean's office at Harvard, so Sutherland used that to buy the optics system. Not only had Bell Helicopter done the technical labour, but the aviation company had

---

* The first head-mounted display was the Stereoscopic Television Apparatus known as Sensorama and built in 1962 by Morton Heilig. It could show 3D films but didn't allow for any interaction. It had smells, fans and movement, so was certainly immersive: one film developed for the Sensorama mimicked riding a motorbike through New York, complete with exhaust fumes blown in the face – as well as, much more appealingly, the scent of pizza. Heilig couldn't find enough investors to continue his work.

also considered how people would interact with this step-removed vision system, via simple physiological experiments. In one, the subject sat inside the building, wearing the head-mounted display and watching a camera feed from the roof of two people playing catch – it's easy to imagine the test subject's head looking back and forth with every throw and catch, like the crowd at a tennis match. 'Then one of the players threw the ball at the camera, and the subject ducked,' says Sutherland in the panel discussion. 'And so it was clear from his point of view, he was at the camera and not sitting in a comfortable office chair.' And that suggested head-mounted displays feel like real vision to users, allowing for a potentially immersive experience.

The 25mm CRTs, designed for the Bell system, stuck out of the side of the headset because it was created for pilots who would be wearing a helmet. It's no wonder Foster looked at the device and wasn't keen to strap it on: photographs show the CRTs and their connecting cables sticking out off the side of the face; the wearer's eyes are visible through small discs of glass, components sit on the forehead and the whole contraption is strapped on over the top of the head. It does not look comfortable. In one image, Foster stands below the 'shower', a four-armed structure that held the cables up, and dangling between two prongs is a tube of cabling that held mechanical position sensors to track the head's turning and tilting – this connected to the strap on top of the headset and is the device that Sutherland called the 'Sword of Damocles', referred to earlier. A wireless version was made, but currents from the aircon messed up the measurements.

Forget hyper-realistic games. This device showed, for example, a simple line drawing of a cube. 'The head-mounted display we built at Harvard was neither very real nor immersive,' Sutherland added. 'You could look at a small volume, you could go around it all, but it was just glowing lines hanging in space. There was no substance to the object.'

Much of the students' work was building hardware to manage the maths to transform the lines, such as to meet changing perspectives or if a piece of a line was off-screen. The trick was doing this maths quickly enough using a computer to draw and redraw the displayed line every 60th of a second.

The computer, fast for its era, wasn't capable of all this, so the team built their own hardware to run the calculations and translate those to coordinates. The head position, measured via the overhead sensors, was pulled into the computer and combined with the line drawing they wanted to display. Next, the data was fed into the matrix multiplier for coordinates, then through hardware to clip and divide the lines, and then pushed through a display driver and amplifier to the headset, so the wearer would see the correct set of lines.

All the hardware the students designed to do maths like clipping and transformation is now built into modern chips. To Bob Sproull, one of the students working on the project, that was part of the charm: this was a purely engineering project, designed to build something simply to see if they could. And they managed it, decades before anyone else.

So when the head-mounted display was finally strapped on to Foster's head, it worked. 'You've got to wonder if you're a little nervous about the design and all that, but it turned out not to be a problem,' says Foster at the VR@50 panel event. However, he admits to inflating the voltage amount throughout the years to make it sound even more dangerous, until he could no longer remember how much it actually was.

Such a project wasn't cheap, so Sutherland and his team of students sought grants – and given the era, a big chunk of the funding once again came from the American defence coffers. When the CIA grant worth $80,000 was leaked to the Harvard 'leftist rag' the *Old Mole*, it sparked a minor scandal and a public debate about whether the military-industrial complex should be meddling with academia. Sutherland didn't speak but did attend the debate, noting at the panel discussion: 'It is evidence of the eloquence of the Harvard faculty or the naivety of Ivan Sutherland that my opinion changed eight times during the debate.'

With that minor controversy over and the project a success, Sutherland left his tenured position at Harvard to join an old friend and colleague, Dave Evans, at the University of Utah, where they ended up setting up their own business, Evans & Sutherland, making simulators for training pilots, which eventually evolved into a digital theatre company.*

But they didn't build VR headsets. 'Many people think that [the HMD project] was the beginning of virtual reality stuff, but I don't think that's the correct way to think about it,' Bob Sproull tells me. 'After this project, virtual reality kind of went nowhere for a long time. There were many reasons, but one is that line drawings are not compelling – they certainly weren't going to do entertainment or games or things like that.'

Sutherland made a virtual reality headset *before* computing was ready, meaning it was necessary to wait for computing to catch up in data processing, graphics and memory. Having made such a system, Sproull told me he isn't surprised by the hurdles faced by those that tried after them – the weight of a headset, the latency delay between the head motion and the

---

* Sutherland and Evans met at the University of Berkeley, California while the former was working for ARPA deciding to set up a company while consulting on a project together. They headquartered the company in Salt Lake City rather than a tech hub for personal reasons: Evans already lived there and had seven children, making a move difficult. They originally wanted to call it the '3D Company', for 'Digital Display Devices', but that name was already taken.

You might not have heard the name David Evans before, but he has a strong legacy in computing science, which goes beyond his own work developing raster computer graphics: his students included Alan Kay, who led the Xerox Palo Alto Research Center; Netscape and Silicon Graphics founder Jim Clark; Adobe co-founder John Warnock; and Edwin Catmull, who co-founded Pixar. Catmull, Clark and Warnock all worked for Evans & Sutherland, too.

corresponding image motion, the limited field of view and the difficulty of providing an image in an eyeglass display are all hard problems to solve. The bigger challenge, says Sproull, is what on earth is the thing even for? He says military and manufacturing come to mind, but is it a consumer product? For gamers perhaps, but for anyone else?

What's telling is that no one in the team chose to keep working on virtual reality headsets. There was certainly no future in them, at least anytime soon.

★ ★ ★

Virtual reality doesn't necessitate a headset; there are other ways to build an immersive, interactive digital space. Myron Krueger explored some examples in the late 1960s, with his Videoplace allowing people in different rooms to interact, such as by tickling; or his Psychic Space: a maze of music and live video displayed on the walls and other surfaces. But when most people think of virtual reality, they don't imagine immersive art experiences, though there are plenty of those popping up so people can interact with artworks by Vincent van Gogh or David Hockney via wall projections. Instead, for most of us, VR is synonymous with headsets designed for gaming. And though Sutherland et al. didn't continue with developing such devices, plenty of others have.

Yet two things happen every time there's a VR boom, from Sutherland's time up to now. First, the technology is massively overhyped as the 'next big thing', and heralded as providing mind-boggling realistic representations of the real world. And second, there is a backlash when it inevitably disappoints.

Let's play a game, no headset required. In what year did Michael Heim, a thoughtful writer about VR, make the claim in his book *The Metaphpysics of Virtual Reality* that computer graphics are now so realistic, with such sharp imagery, natural shaded textures and light radiosity, that they bring to mind the idea of virtual reality? Was it 1993, 2003 or 2013? My setup probably reveals it's the earliest date – and if you're old enough to have played a computer game, VR or otherwise, from that decade, you may be wondering what Heim was on about. There was no chance of mistaking computer-made graphics for actual reality back then; the biggest games that year were a sequel to *Street Fighter* and *Super Mario Kart*.

But compared to what came before – think of Sutherland's line drawings of cubes – the pixelated drawings of that era understandably felt like an unreal leap into the future. Indeed, given the computing power of the time, it would have been difficult to imagine how graphics could have got better, but thanks to innovations like GPUs and rendering techniques, we now have computer-made graphics that are incredibly photorealistic – we can fill in the face of a dead actor, such as Peter Cushing, who appeared in *Star Wars: Rogue One* 22 years after his death.

Successful virtual reality shows that graphics quality isn't the most important feature. Immersion requires a wide enough field of view – how far you can see around you – while eye tracking, latency and optical quality are all key to immersion, as any distortion can ruin the effect and cause motion sickness. If you nail those requirements, most people will be happy walking around in a clearly cartoonish world.

At least I was.

Zero Latency is a VR gaming company. You and your friends gather in a massive room – think warehouse-sized. The space must be expansive to allow players to walk around with the headset on (connected to a backpack-carried computer), seeing not the empty room but computer graphics. One of the tasks is to walk along a path that starts curving above your head, so it seems as though you're walking upside down – and it *does* feel that way. So much so that I had to hold my friend's hand, knowing full well I was just wandering a warehouse wearing a weird headset, and not actually defying gravity by strolling around the loop of a rollercoaster. Eventually, with much moral support, I managed that task, and we went on to play a spaceship-themed shooting game, and you'll not be surprised to learn that I 'died' constantly.

The graphics weren't mind-blowing; the digital set felt like low-budget sci-fi. There was no belief that this was a real spaceship. But the game felt wholly immersive thanks to the interactivity, the good range of vision, and the ability to move and look around. The boxes I ducked behind looked like computer graphics, but I ducked behind them all the same. I knew the spaceship door wasn't real, but I still waited for it to open before stepping through.

Graphics quality can be questionable, but the tech *does* matter. The headset can't be too heavy, it must react quickly to movement and the software must be engaging for these experiences to trigger the right parts of our brains. If any part of that recipe is missing, people can't wear it for very long, get sick or, worse, bored.

Back in the 1980s, 20 or so years on from Sutherland's work, the idea of virtual reality remained tantalising, but the graphics, the experience and the hardware were decidedly not quite there yet – though plenty of people around the world were trying to make it happen. In 1978, researchers at MIT built an explorable virtual tour of ski town Aspen, Colorado by recording video while driving around, a project not terribly far off Google Street View. Around the same time, artist David Em worked on navigable worlds at NASA's Jet Propulsion Laboratory. You could immerse yourself in virtual worlds if you only had the hardware.

And hardware was already in the works, at a location near Boston, Massachusetts, where Eric Howlett was fiddling with stereoscopic lenses, which take images from two different angles to add depth. Howlett's optical system LEEP (large expanse, extra perspective) offered a wider field of view than existing techniques, and he was granted a patent in 1983. He didn't intend it for VR, but for photography, hoping Polaroid or Kodak would show an interest – they didn't. Instead, Howlett tried building cameras independently: he bought the necessary components for 70, managed to build 20 and shipped 'the only three cameras we could get to work properly', he said. Of those, one was stolen and the other two required repairs by their 'ingenious owners'.

At this point, Howlett was already in his fifties having had a career that ranged from developing radar to running a mail-order business and eventually manufacturing displays. His son Alex, with whom he founded LeepVR, was born when Howlett was 57. Developing the technology together helped them to feel closer despite the generational divide.

By the mid-1980s, Howlett was running low on cash. But luck struck, and the core stereoscopic viewer attracted the attention of two early pioneers of VR: NASA wanted it for its digital office, the Virtual Interface Environment Workstation (VIEW); VPL Research, the company behind products such as the EyePhone and DataGlove, wanted it too.

Michael McGreevy, who was the project lead for the Virtual Visual Environment Display (VIVED) headset and VIEW at NASA's Ames lab began ordering viewers. He also advised VPL, who NASA had been supporting, to use the LEEP lenses. NASA and VPL were the two leaders in 1980s VR, linked not only by location, just a few miles apart in Silicon Valley, but also by people. Those people had become available to be snapped up by VPL and Ames after a gaming failure.

In the 1970s, amid the boom in video and arcade games, Atari – maker of Pong – was considered one of the fastest growing companies in the US, had $2 billion in sales. It was acquired by Warner Communications in 1976. Like any good tech giant, Atari spent some of that cash quietly setting up a secretive lab. And in 1982, its newly founded Research Lab was building VR before the term was even in use.

Its developers started by working on an arcade game that mimicked Morton Heilig's Sensorama, an early immersive arcade-style experience. The user would sit on a chair and put their head into a box to experience 3D images, sounds and even smells, capturing what it was like to cycle through New York City. Atari's aim was to haul that analogue idea into the digital world and make it interactive.

To build such a machine, Atari had splashed its cash, hiring the best minds in the nascent business. Project leader Scott Fisher was poached from the MIT Architecture Machine Group (which eventually evolved into the MIT Media Lab) after a visit from the Atari Research Lab's director Alan Kay, who was himself poached from the famous Xerox Parc labs. Kay was so flush with cash and so keen to hire Fisher that he even offered to buy the MIT lab, though he naturally failed (not everything is for sale); instead, Fisher and his coworkers were offered serious money to shift jobs. 'There was a lot of money. It was just a big sandbox,' said Fisher in an interview.* '[We were] thinking about stuff 20 years down the road. That was kind of the holy grail – blue sky stuff.'

As fun as those few years were, VR built by Atari was not to be. The 'Atari shock' hit in 1983: the video game market crashed – blamed on everything from market saturation to competition from PCs – and Atari closed the lab. And that meant there was suddenly an oversupply of extremely intelligent people with very specific skill sets and large corporate redundancy payouts. Some of them teamed up to build an industry. They included Jaron Lanier and Tom Zimmerman, who went on to set up VPL Research; Scott Fisher, who went on to NASA's Ames lab; Brenda Laurel, who co-founded Telepresence Research with Fisher and is considered a trailblazer in game design, and Michael Naimark, who worked on that interactive map of Aspen and later with Google, Apple and LucasFilm on virtual projects.

* This is from an article at KillScreen about the secret history of Atari: https://bit.ly/485rxqc

The first-named of those figures, Jaron Lanier, is a key person here, and he often gets the 'father of virtual reality' title that the media (again, sorry) loves to bestow, in part because he's credited with coining the phrase 'virtual reality', though he himself disputes this.

Lanier is what is known as a character: with trademark dreadlocks and a collection of obscure string instruments, he grew up in New Mexico in a house partially made of geodesic domes that his father let him design after the sudden death of Lanier's mother in a car crash and after their previous house burned down.

An unquestionable prodigy, he learned programming while taking college courses as a young teenager, and that's where he read Sutherland's 'Ultimate Display' paper. He recalls in his memoir: that he felt overwhelemed reading about Sutherland's work, perhaps more so because he couldn't see a demo or film of the technology in action, and instead imagined it all.

Not one to follow a traditional path, Lanier spent time amid New York's avant-garde music scene and worked as a midwife's assistant. When his enthusiasm for virtual worlds brought him back to Silicon Valley, he assumed he'd struggle to find a meaningful job, having plenty of useful skills but little in the way of traditional work on his résumé. But instead, his first interview was with a gaming outfit put together by George Lucas, hot off the success of *Star Wars*. What's the one way to blow an interview at *Star Wars*? Tell the suit asking you questions that you prefer *Star Trek* because Gene Roddenberry believed people would become nicer as technology evolved, which was more interesting.

He didn't get that job. Lanier had no regrets and eventually ended up at Atari, programming and designing games, before being asked to join the Atari Research Lab, based in Massachusetts. There he met the great Marvin Minsky as well as future collaborator Tom Zimmerman, so he didn't take it too hard when the Atari shock cost him a job. Lanier took his games royalty payments with him back to California, to a cheap, uninsulated hut in Palo Alto.*

There, he and Zimmerman began cobbling together the hardware leading to his startup VPL Research, which included the interactive DataGlove to track hand gestures, eventually selling this as a gaming accessory to Mattel and as a robotics controller to NASA. The DataGlove paired with a VPL headset called the EyePhone – yes, that's really what it was called. The EyePhone was the first commercial VR headset, and modern models still look rather like it, though are lighter than this groundbreaking predecessor, which infamously left red marks on the faces of people who demoed it. Weight aside, Lanier says the EyePhone had a resolution and field of view that was as good as any today – but the headsets cost $50,000 each. It was simply too expensive, and with the nascent VR market floundering, VPL eventually filed for bankruptcy, with its patent portfolio later snapped up by Sun Microsystems after the mid-1990s VR crash.

But that's getting ahead of ourselves. Back in the 1980s, further down the El Camino expressway that runs through Silicon Valley was NASA's Ames Research Institute in Mountain View, a moment away from Google's main campus. That's where Scott Fisher turned up when Atari Research imploded.

In the mid-1980s, Ames was working on VR research, helped by VPL. In 1986, Michael McGreevy set up a project to build a headset known as the Virtual Visual Environment Display (VIVED), which used LEEP's stereoscopic optics with LCD screens, with a 120-degree field of vision, head tracking and even speech recognition – all the key features of

* Things have changed. The Silicon Valley city of Palo Alto is one of the most expensive for real estate across the US, with an average house price above $3 million. If you want a cheaper home, head down the main drag El Camino Real to Sunnyvale or Mountain View, where houses have a much more reasonable average asking price of $1.8 million.

a modern headset. This could be paired with a gesture capture system like the DataGlove.

NASA intended VIVED to be used for controlling robots and other hardware in space, and as such, the headset looks like an astronaut helmet. A separate, lightweight workstation-mounted viewer – think a high-tech version of the red plastic View-Master* toy that lets children click through a ring of different images – was also built for information-management tasks and for 3D interaction with your files. This provided a 'moveable window into a three-dimensional data space'. Sounds more fun than clicking through spreadsheets, at least.

NASA handily released a video† of what it looked like – and it's both impressive and hilarious. A man wears a massive headset with the NASA logo across the front and a thick cable running off the back, paired with a black set of DataGloves from VPL on his hands. The office setting fades to a boxy digital representation, just grey blocks where computers and books once were. A virtual hand enters the view, matching the gloved hand, moving disembodied through the greyscale place and navigating a menu that drops down from nowhere, letting the headset wearer select 'toggle flying'.

A robotic voice intones: 'FLYING ENABLED.' And then it gets *weird*. The view dramatically zooms out into a high overview angle, the office disappearing. A series of doors appear, looking as if drawn by a child, with the words

* Fun fact about the View-Master: it was designed in 1939 by William Gruber and his business partner Edwin Mayer, but not as a toy. Rather, it was intended as an alternative to postcards, with the idea that visitors would buy the reels of photos from gift shops as souvenirs. The stereoscopic devices were used by the US military for training in the Second World War, and in the 1960s reels were sold promoting television shows, before the gadget finally evolved into the toy it's seen as today.

† You can watch the video – and you should – on YouTube here: https://youtu.be/H0EI6KLnnSE

'ENTER'. Game-like music plays, and the user can interact with various digital 3D shapes that play different sounds.

In the next scene, the headset-wearing person uses the DataGlove to control a robot and play a virtual instrument. The video ends with the user flailing his arms to bash the virtual drums, as his colleagues filter into the room and take down black curtains hung over computer racks. Laughing, they grab his arms and then pull his headset off, as though he was lost in VR and they rescued him – gosh, smart people are goofy. All of this developed the technology that is still at the core of the VR systems we use today, but there was still a lot of work to be done – details really matter when the technology is strapped to your head.

* * *

Not everything happens in Silicon Valley, believe it or not. Forget California, and picture rainy Leicester in the UK. In the 1980s, Jonathan Waldern was working on a PhD in 3D interactive graphics at Loughborough University, where a prototype workstation he built won a small prize from telecoms firm BT. That sparked something, and after finishing his studies Waldern mimicked the greats of Palo Alto and set up shop in his home garage in order to build a VR system in his off-hours from working at a local computer company. When casino and restaurant entertainment company Leading Leisure coughed up £600,000 in return for a 75 per cent stake, Waldern was able to move W Industries was able to move from the garage to an old mill.

By 1991, W Industries had been renamed Virtuality, and Waldern had sold 390 virtual reality systems – usually seated systems for arcades running games like *Dactyl Nightmare*, in which giant pterodactyls were hunted with crossbows – across mostly the US and Japan, though Waldern believed the technology had better uses. 'We had no alternative. The leisure industry could see its application. They put up the money, so that's what we've produced,' he told an interviewer in 1993. 'But its uses in other fields are so obvious.' He pointed to architecture, surgery (for which the technology was being tested by the University of Milan, he said) and even sex, with pornography firms approaching Waldern (he turned them down, despite the potential revenue). That might have helped save the company. But Virtuality largely relied on coin-operated arcades, and though its games were astonishing for the era, the cost of £4 for three minutes of play was a lot back in the early 1990s.

★ ★ ★

The first boom time for VR happened in the 1990s. But it ended not in an explosion in VR and gaming, but an absolute implosion that likely set back the industry.

Sega VR is the most famous example of vapourware – a product that was announced but was never actually released. Announced in 1991, the aim was brilliant: take all those overpriced coin-op VR games by companies like Virtuality and bring them into the home.

Sega did make hardware that was actually seen by 1993, shown off at the Consumer Electronics Show (CES). 'This new device takes us into the future, the future being virtual reality – something that's being talked about by other people but produced by Sega,' said MTV presenter Alan Hunter, hired to unveil the device. Plenty of media coverage followed: Sega VR made the cover of *Popular Science*,* although the story admitted that so far developers of VR seemed to outnumber customers. Still, it was exciting: consumer headsets were arriving alongside predictions of more mainstream, affordable headsets to follow. Amid the rise of VR games from Virtuality in arcades and shopping malls

* Also in that issue: heat pumps going mainstream, scheduling your appliances to save on energy use and tools for working out of the office – because nothing at all is new.

everywhere, Lanier was plastered across the cover of serious magazines, while at the cinema, the 1992 film *The Lawnmower Man* even featured the VPL EyePhone as a prop.

Delivering a VR headset to match those hyped expectations was always going to be difficult, and sure enough, Sega was struggling to pull all the necessary pieces together at the promised price point – though the company did manage to solve one key problem, managing to source the necessary sensors for head tracking from a company called Ono-Sendai for $1 per headset (though this is somewhat disputed, it was in the dollars range at least).

A stickier problem couldn't be fixed: nausea. The Sega VR gave some people who tested it motion sickness, a challenge still faced by modern VR and AR manufacturers. That was reason enough to halt the release, but news reports also claimed the way the optics were set up meant it was possible to adjust the settings in such a way that viewers' eyes were forced to look just a bit outward in opposite directions, which could easily cause eye damage, a liability the company's lawyers weren't willing to overlook. Sega itself said the headset was simply so good that people would forget they weren't in the real world and hurt themselves. Sure, that could be it too. Whatever the specific cause, despite years of development and after building up oodles of positive marketing, Sega shelved the project and Sega VR remained vapourware.

Into that eager enthusiasm-already-edging-into-disappointment, Nintendo lobbed Virtual Boy. This headset-style system was mounted on a non-adjustable tabletop tripod with displays that only used black and red. At its core, the Virtual Boy employed a stereoscopic technology by Reflection Technologies Inc. (RTI) that used a 2.5cm (1in) monochrome display paired with red lights and stereoscopic optics to produce a 3D image. The head of RTI, Allen Becker, pitched it to toymakers, including Mattel, who turned him down, before trying Sega – which also turned him down. That's no particular coincidence, as Tom Kalinske was the CEO of Mattel before moving to Sega, meaning Becker had to give the same failed pitch to the same unimpressed CEO. He had better luck at Nintendo, which was looking for a new gaming system to invest in.

After Sega's failed headset and the rumoured reasons that led to it, it's perhaps no surprise that Nintendo decided to mount the Virtual Boy on a tripod so players had to stand or sit still while using it. Gunpei Yokoi, the hardware lead at Nintendo who created the delightful Game Boy, gave an interview in 1994 admitting safety concerns but also explaining that the company was considering making the headset more mobile with a shoulder mount, so players could use it without a table.

Even if the weird red colour and lack of support for Nintendo's massive back catalogue of games, including key titles like Mario, weren't enough to doom this product, concerns over motion sickness were so severe the game recommended breaks every 15 minutes – no surprise then when it was pulled from sale after only a year.

The Virtual Boy sucked so hard that the VR industry imploded,* and it was another 20 years before anyone dared go big with a mainstream headset release.

★ ★ ★

But in 2010, a teenager living in a trailer in his parents' front drive managed to perfectly time his foray into VR headsets. After setting up a massive six-monitor rig to make gaming feel more immersive, Palmer Luckey realised it made more

* Of course, people just couldn't stop trying, even despite the evidence in front of their eyes. In 1999, Philip Rosedale formed Linden Lab to design VR hardware, making a pre-commercial 'The Rig', which players could wear on their shoulders. Eventually he gave up on the hardware and shifted into virtual worlds, building the multi-user online world *Second Life*.

sense to have a headset. He started buying up the niche headsets that were still being made outside the mainstream, collecting 50 different models, which he largely paid for via modifying PCs and repairing iPhones.

That collection taught him everything he needed to know about VR's previous failures. Headsets were hilariously limited, making it appear like you were looking through toilet paper tubes. Luckey – a kid in an odd living situation who finished school early and took university courses for fun, sounding remarkably like Lanier (though in other ways, very much not) – spotted this core challenge and solved it on his own. (Another problem, motion sickness, took a bit longer to solve.) In 2010, he unveiled the very first prototype to a VR forum. PR1, as he dubbed it before renaming it the Oculus Rift, lacked 3D and was excessively heavy, but the large display offered a 90-degree field of vision, well above rival headsets. He kept at it, figuring out simple solutions for technical challenges: why use two screens for 3D when you can just share one screen for both eyes? Why shell out for high-end optics when you can just fix distortion in the software? Those smart kludges helped keep costs down.

At this point, he planned to sell the headset as a DIY kit online. 'I won't make a penny of profit off this project,' he wrote on the forum. 'The goal is to pay for the costs of parts, manufacturing, shipping, and credit card/Kickstarter fees with about $10 left over for a celebratory pizza and beer.'

Lurking on that VR forum was legendary developer John Carmack, who made *Quake* and *Doom*. Carmack messaged Luckey, asking if he could see a prototype. Starstruck, Luckey shipped it immediately, and Carmack took it to the E3 games conference, saying it was the best VR he'd ever seen. Like sharks, investors started circling. But an online buddy introduced Luckey to Brendan Iribe, who had games credentials as a designer on *Civilization IV* but also business experience via his user interface company Scaleform, which sold to Autodesk for $36 million. Iribe brought along his colleagues-cum-friends Michael Antonov and Nate Mitchell to help coax Luckey through the business side of the project.*

Luckey knew he still wanted to crowdfund production on Kickstarter, hoping for a hundred or so sales to games developers – a worthy and reasonable goal. His new investor pals believed he should ask for $500,000 and rethink the campaign as sales rather than dev kits. They split the difference on $250,000, but it didn't matter much in the end, as they surpassed that goal in less than two hours. Ultimately, the Kickstarter raised more than $2.4 million from 9,522 backers, even though the $300 headset was still described as a developer kit.

The Oculus team didn't just ship what it had, it continued to refine the product. Unlike Sega, giving up when faced with challenges like motion sickness, the team did something new: it tried to solve them. Some solutions were in the hardware, with lower-latency AMOLED displays and additional head-tracking sensors. Others were in the software, including an algorithm that predicted player movement so that images could be preloaded. The first version of the headset came out in early 2013, followed by an updated version – but Oculus never actually released a commercial product on its own, because in stepped Facebook (now Meta) with a $2 billion acquisition. Luckey was only 21 years old, and it was just three years after he'd finished that prototype while living in a trailer on his parents' driveway. He reportedly earned $500 million from the deal – and he immediately bought a 1969 Ford Mustang, a Tesla Model S and a helicopter. Treat yourself, after all.

With that, kaboom! VR exploded. The same year Facebook took over Oculus, Valve showed off its SteamSight prototype. The next year brought Samsung's Gear VR, followed by

* Blake Harris' book *The History of the Future: Oculus, Facebook, and the Revolution that Swept Virtual Reality* includes a detailed telling of how the group came together to make the Rift happen.

HTC's Vive and Sony's PlayStation VR a year later. VR headsets had arrived.

For the most part, these headsets worked in similar ways to the Oculus Rift. They each had a small display, like a miniature of the monitor on your desk, that you view through stereoscopic lenses, which distort the view to make it seem bigger and further away, adding 3D depth. The view shifts when you move your eyes or your head, thanks to trackers for both. The headsets of course need processing power, which can be onboard or on a tethered PC or console.

There are different ways of achieving all of the above and the way it's all pulled together matters. Key specs include the resolution of that tiny display; the higher the numbers, the better the image quality, but that will come at a higher cost and require more processing power and battery life. Higher refresh rates, referring to how often the image changes per second, can help reduce motion sickness while improving animation quality, but again there's a cost and processing trade-off.* One problem Palmer solved was the field of view; the wider this is, the more natural it will feel – human eyes have a horizontal field of view of 200 degrees or so, and most headsets have half that but are immersive enough to convince players.

What was this first true generation of VR headsets like? I had a go with a PlayStation VR at the London Science Museum's launch of its Power Up games exhibit. Not a natural gamer, I started with *Tetris*; the whooshing stars alongside the simple puzzle made me instantly nauseous. I pulled the headset off, my steamed-up glasses falling from my sweaty face. After a break, I tried *Wipeout,* a spaceship racing game that made me sick for the first few laps, but I stuck with it, my piloting improved and it became less nauseating – it's clear

---

* We humans can handle very low frame rates and much faster ones, but anything between four to 12 frames per second causes confusion – what SRI's Thomas Piantanida called the 'Barfogenic Zone'.

I'm as much the problem as the VR (I do get car sick easily.) Once the sick feeling eased, it was clear why people like this view of gaming: the immersion fades the rest of the world to nothing and all that matters is getting the spaceship car around the track. I continued to do this – badly, cursing as I went – until someone bumped my chair and I recalled that I wasn't on my own but very much in public.

Powered by Oculus hardware, the Samsung Gear VR was a bit different than the others, as it used a Samsung Galaxy smartphone as the display and computer, while the VR unit had the optical bits and pieces; this was possible because of Luckey's clever idea to use a single display split into two to make stereoscopic 3D. Costing just $99, they were given away with purchases of high-end Samsung Galaxy smartphones. Google also stepped back into the VR world with Cardboard, a basic cardboard viewer that holds your smartphone, housing simple lenses. Cardboard-compatible apps would split the view into two to make a stereoscopic 3D image. There were games, apps for meditation and reading the *New York Times*, and a version of Street View, plus art tools built by Michael Naimark, one of the group of VR overachievers from Atari. Despite the cheap price, the devices didn't take off.

VR wasn't winning for everyone. Run by Facebook, Oculus struggled to make headway, with sales estimated at just 400,000 Rift units in 2016, fewer than rivals, with the Sony PlayStation VR shipping more than a million – the $2bn deal was deemed one of Zuckerberg's 'rare mistakes'.* The same year, Facebook gave Luckey the boot amid controversy that he reportedly funded an anti-Hilary Clinton group. 'Selling Oculus to Facebook was the best thing that ever happened to the VR industry, even if it wasn't super great for me,' he said later. He's since gone on to start a military drone company, and – surely with a nod to

* In hindsight, Zuckerberg has made many, many mistakes.

Sutherland's paper 'The Ultimate Display', which suggested virtual deaths could feel real – blogged about a design for a headset called NerveGear that would fatally smack a person in the skull if they were shot in-game. 'The good news is that we are halfway to making a true NerveGear. The bad news is that so far, I have only figured out the half that kills you,' Luckey wrote. He may have lost his company, but not his sense of humour.

Oculus had unquestionably rebooted VR, but sales weren't as robust as hoped. Startups were selling for billions. You could try VR with a piece of cardboard. Truly, the VR revolution was here. But sales of VR headsets have remained a bit of a disappointment, with only 10 million units shipped globally each year. For comparison, Nintendo sold 19 million Switch hand-held game consoles in 2022, while Apple ships at least 10 million iPads each *quarter*.

Why? There's the physical side: setting aside headaches and nausea, perhaps the appeal of wearing a headset while gaming is limited, as it isolates the player, removing any social component and pinning them in one place so they don't trip on living room furniture. And beyond games, no mainstream, entertainment applications have ever really materialised. There have long been suggestions that we'll travel in VR rather than hop on a plane; that we'll watch movies, learn or even meditate. But who wants to wear a headset while practising mindfulness? What school can afford a headset for every student? And who would rather look at pictures of the beach than be there? Sure, you don't have to worry about sand getting everywhere, but we go to feel the sunshine, smell the sea air and drink margaritas, not just look at them.

But beyond gaming there are niche uses that have real value, from training to PTSD treatment, with researchers like Skip Rizzo at the University of Southern California using VR for more than a decade to treat US soldiers returning from warzones and even victims of the 2015 terror attack in Paris.

Even people who own VR systems don't use them all that much, with one study suggesting headsets are used an average of six hours a month, with a quarter not touched for three months. Can AR best that? VR headsets are another way that technology cuts us off from the world – hence the appeal of augmented or mixed reality. Rather than fully immerse yourself, augmented reality or mixed reality layer digital bits over the real world, letting the virtual and real worlds be woven together. It's the end of screen time, but the beginning of always looking at a screen. And all you need to do is wear glasses, whether you need them or not.

★ ★ ★

And that's what Google was trying to create with Glass in 2012. It all started about two years before Brin announced the AR glasses in that hyperactive demo, and perhaps unsurprisingly, the germ of the idea sprung into life amid a high-speed sport. Sebastian Thrun – whom you'll recall as the researcher who won a DARPA driverless car challenge before helping found Google's X labs – and Google's two co-founders, Larry Page and Sergei Brin, often went skiing together. What did they talk about on the chairlift up the hill? 'The future of everything,' Thrun tells me.

That included a long-running conversation about shrinking computers, and how reducing the size of devices made computing more pervasive and closer to users, shifting them from monolithic mainframes to PCs that stayed on desktops, to laptops we ferried back and forth on our commutes, and then smartphones that never leave our sides. Though smartphones have allowed for truly ubiquitous computing, there is still a gap between user and device. 'You have to divert your attention,' Thrun says. 'You had to take something out of your pocket to stare at it instead of the world. It wasn't quite the seamless integration of the physical world and virtual world that we wanted.'

At first, Thrun thought brain implants might be the answer, and plenty of people still do. But Thrun also realised it might be a tough sell 'to go to a customer and say: hey we have to open your skull'. Instead, the trio came up with the idea of putting the sensors right on your face, letting this next-gen device see and hear what you do, making it possible to just talk to it as an interface.

It took some work. At first, Thrun's team took a mobile phone and attached it to the side of the head – picture your phone strapped to your head and you have imagined it perfectly. From there, they simply picked and chose the necessary components – a less demanding processor and a battery to match, for example – and placed them along the arm of the glasses, some in front of the ear and some behind. 'This thing was super easy to wear,' says Thrun.

But though Glass failed, the team managed to shove a computer with Wi-Fi, speakers, a display and a battery into a package that weighed 42g (1½oz) and could sit on someone's face relatively comfortably. 'That blew my mind, that we could do this,' Thrun says.

So if the hardware worked, what happened? 'It should have been released later,' says Thrun. 'For whatever reason, we released it way too early and the software wasn't up to snuff... That was probably the single biggest mistake.' Glass struggled to find consumer sales and the project was promptly shut down, though briefly revived for business use – it's handy to get directions, assembly instructions or repair advice in a hands-free way.

There were other mistakes that Thrun sees in retrospect. Glass was marketed as a business productivity and lifestyle device when it should have been pushed as an outdoor sports activity device. After all, it was dreamed up on a ski hill. That would have helped avoid the privacy issues inherent to wearing a camera on your face, as Glass would have been seen as a personal sports device rather than something to wear to the bar, or, as one technology journalist did, in the shower.

Plus, it would have helped on a technical level to make Glass into sunglasses, as the polarised light made the display look brighter. The last issue that Thrun identifies is the name, partially because of the 'glassholes' moniker that was quickly bestowed on users of the technology. 'None of us saw this,' he says. 'We had a world-class agency doing the naming.'

Magic Leap had a better name, but it wasn't much help. Founded in 2010 by Rony Abovitz, who had previously founded a surgical robot company, this mixed-reality headset company followed the pattern set by Google Glass in a few ways: massive hype, disappointing entry into the real world and pivot to business.

First, a bit about Abovitz. The hype around Magic Leap sparked curiosity about the little-known entrepreneur, and it didn't take much googling to find a lot to be curious about. Alongside running a company valued in the billions, Abovitz played in an indie-rock band called SparkyDog and Friends, ran a blog full of tech philosophy theories – 'if [software] can wirelessly transmit from BlackBerry to iPad to PC, why can't our souls float as they will?' – and had given a bewildering TedX talk in Florida while dressed as an astronaut. Creative minds can be awfully, well, creative, but he was correct on some points, blogging this in 2005: 'Thanks to Google and blogging, every little bit of insignificant data will be embalmed forever in some server, haunting you.' Not wrong there.

Back to the technology. Magic Leap released a video while still in stealth mode, showing a CGI whale leaping up into a school gymnasium. As it turned out, that image was just a marketing mockup. A video demo Magic Leap released of 'a game we're playing around the office right now' showed a first-person shooter overlaid around and over their desks, but turned out to be nothing but a fake concept pulled together by a movie studio.

With the technology heralded as the arrival of true AR, investors gathered like a troop of monkeys spotting a cartload of bananas, and they really made macaques of themselves, driving the value of Magic Leap skywards. In five years, despite not releasing a product of any sort, the company managed to raise $2.6 billion in investment, nab Sundar Pichai, later Google's CEO, as a board member and hire sci-fi author Neal Stephenson as its 'chief futurist' – he turned the ultimate failure of the project into an audio drama called *New Found Land*.

How could so many people have been convinced by Magic Leap? There were two versions of the technology: one dubbed 'the Beast' and another smaller, wearable prototype called 'the Cheesehead'. Tech site The Information revealed fairly early on that the mind-blowing, wallet-opening version – the Beast – weighed 'several hundred pounds' and used a motorised projector to create images.* That was what Magic Leap showed off during its first round of financing, though there was also a pair of sunglasses attached to a battery pack on the table, a suggestion of what the final product would look like. In 2018, Magic Leap did release a smaller $2,300 device, the Magic Leap One. It did not go well. One investor told a Bloomberg reporter: 'The first time I put on the real one, it was just like, "Oh shit. You guys did not deliver on your promise."' Like other AR, it was like 'looking through a large floating window', noted one review, with *The Verge* calling it 'flawed'. Palmer Luckey took one apart alongside tech dissectors at iFixit to see how it worked, declaring it 'tragic', with the core, supposedly secret, technology consisting of 'photonic lightfield chips', nothing more than the 'same technology everyone else has been using for years'. Sales reflected such negative reviews, sparking mass layoffs and a pivot to business customers – as well as a new CEO.

That wasn't enough to scare off others from diving into the AR market. Snap, the makers of the SnapChat app, released glasses in 2016 that let you record short videos and upload them via your smartphone connection, the technology largely driven

---

* You can read the full story on The Information here: https://biturl.top/vYJFJn

by the acquisition of Vergence Labs, which had made its own point-of-view video glasses three years before. A next-gen model that's still in the works promises to 'bring augmented reality to life' by overlaying computing onto the world.

And in 2016, Microsoft unveiled HoloLens, a head-mounted mixed reality platform for $3,000, followed by a fully supported corporate version the next year for $5,000. When HoloLens was announced, the *New York Times* called it a 'sensational vision of the PC's future', but six months later rowed back on that, saying a recent demo offered a 'reality check' and that AR has a 'way to go before the technology is ready for the masses'.

That was partially down to the limited field of view, described as a 'small rectangle suspended several feet in front of your eyes'. Industry players like Asus, Acer, Dell, HP and Lenovo all signed up to develop headsets with VR capabilities using HoloLens gadgetry, with the aim of selling the hardware for prices as low as $299. Not a single one was ever released, though.

However, HoloLens has been used for teaching medical students, boosting efficiency in factories, and Microsoft even signed a deal with the US military, sparking a small protest from employees. Microsoft kept the deal, but reports leaked out suggesting that the Integrated Visual Augmentation System goggles, which are specialised versions of HoloLens, were making soldiers sick.

After acquiring Oculus, Meta, the company formerly known as Facebook, has been churning out devices ranging from a cheaper VR version to smart glasses, with a high-end augmented-reality headset, the Quest Pro, selling for $1,499. Meta, according to *The Verge*, plans to release more fully featured smart AR glasses in 2027, after a decade in development.

* * *

There's one easy way to judge if a technology is ready for the market: does Apple make one? Apple has been working on an augmented reality headset since 2015 or so, indicated by a series of acquisitions and hires, including German AR firm Metaio in 2015 and Canadian headset maker Vrvana in 2017, as well as former NASA employee Jeff Norris in 2017.

After years of such teasing, Apple made it official, unveiling in June 2023 the Apple Vision Pro for $3,499. Apple is calling the experience 'spatial computing',* which is effectively augmented reality. You can see what's going on around you (such as in your living room), with digital images overlaid on top.

The ski-goggle-style headset is controlled via voice and hand gestures, as well as eye tracking, letting users navigate through apps such as viewing photos, and take video calls using an uncanny-valley digital avatar of yourself.You can use it for work, wearing it instead of setting up multiple monitors, or game on the equivalent of a 30m (100ft)-wide display. One innovation, if you can call it that, is a display on the front of the goggles that shows a live video feed of your eyes, to make interacting with non-goggled people feel less strange. I'm not sure a video feed of eyes makes anything less strange, though.

While Google was eager to have everyone on stage wearing Glass, *New York Times* reporter Vanessa Friedman noted that no one actually wore a Vision Pro during the launch demo, suggesting Apple was well aware not only how silly it looked, particularly the cable running down to the battery pack, but also that they risked become memes online.†

It's also interesting what Apple didn't (yet) build: AR glasses. The dream of pulling on a pair of glasses and getting notifications or playing games without glancing at your phone isn't yet feasible, based on what the world's most successful technology maker has managed to produce, at least. Naturally,

* CEO Tim Cook apparently finds Facebook's favoured term 'metaverse' offputting to average people; and they go with *that*?

† You can read her full article here: https://biturl.top/6beQVf

Apple hasn't officially said anything on the subject, but reports from Bloomberg suggest work on lightweight AR glasses has been delayed, perhaps indefinitely, owing to technical challenges. Apparently, the plan was to release both the Vision Pro and the AR glasses simultaneously, but the latter may now take years to come to market – if they make it at all.

* * *

Will the hype of AR glasses, Apple or Meta or otherwise, one day live up to the reality? Probably – these are engineering challenges, and with a few more Jaron Laniers or Palmer Luckeys, perhaps they're the solvable kind.

But it remains to be seen whether we want to see a digital overlay in front of our eyes as we go about our days. For the answer is … yeah, probably. If you'd asked someone in the 1970s if they'd carry around a small glass rectangle and spend most of their day staring at it, they likely would have focused on the downsides – what about battery life? why carry it in your pocket all the time? who even needs constant updates? – yet here we are and it's the norm.[*]

A different question is whether that means the so-called metaverse is the future. The metaverse is simply the virtual side of all this hardware, where those digital avatars do whatever it is they do. Perhaps you send your avatar off to a meeting, or a Beyoncé concert where you 'sit' next to a long-distance friend and enjoy the same viewpoint, or, as Mark Zuckerberg has suggested, hang out and play cards with the digital representations of your family. To be clear, you're still

[*] This isn't an assumption I make from my privileged position in a wealthy, Western country. There are almost as many smartphones in the world as there are people, though of course many people do carry multiples (CEOs, drug dealers, etc.). While smartphone penetration is about eight in every 10 people in North America and Europe, it was about 64 per cent in Sub-Saharan Africa as of 2021. Smartphones have truly taken over the world.

doing all these things, but rather than you sitting there physically, your avatar takes your place in the virtual world.

Some of that has appeal. I have friends and family scattered around the globe, but when the COVID-19 pandemic sparked lockdowns, virtual games, Zoom parties and online shared experiences became the only way to hang out. People like to point to Travis Scott's *Fortnite* concert, which for the uninitiated is a musician playing a concert inside a video game, drawing 12 million viewers, but that's an exception rather than how gigs are done now.

That said, virtual gigs were the one digital lockdown activity I didn't hate. Unable to travel and perform at live gigs, my favourite singer, Dan Mangan, made use of an online platform he'd founded for booking independent artists to offer live-streamed shows. It was him singing in his basement, but my Canadian friends and my own family would watch them simultaneously, saying hi to each other in the text chat. It was charming and sweet – and was done over video conferencing, so no headset was required.

Stretch the idea to a headset, and perhaps you could watch a football match with your friend from the same vantage point, talking to each other as if you're both there, sharing the same experience, virtual though it may be. Would that AR experience be better or worse than watching the game on TV while texting? I'm not sure. Keep in mind that the metaverse version requires an expensive headset that you have to actually wear, isolating you from anyone else on your sofa, regardless of whether there's a live stream of your eyeballs on the front. AR can help that, letting us see both the real world and the virtual, but augmented or mixed reality needs to look more like the hologram promised by sci-fi than a broken TV off in the distance.

And, with the shift to home working, there's the possibility of using such headsets for our jobs. Apple calls this 'copresence', believing two remote colleagues will feel like they're working together. This could be truly useful in bringing together experts on opposite sides of the world to solve a unique and tricky problem. But it's a high cost to pay to make Zoom feel a bit fresher.

And it has been a high cost. Companies like Sega likely shelled out millions to make a product they couldn't bring to market because it made people sick, though that didn't stop the company from letting its hype machine go into high gear. Investors blew $3.5 billion on Magic Leap. Meta spent $2 billion buying Oculus when it didn't have a product yet, meaning it had to fund the rest of the work. Apple has reportedly spent $20 billion over the years – though that's of course not confirmed – and still had to limit its release to the Vision Pro, rather than the hoped-for smart glasses. That's the thing about VR and AR headsets: people spend a lot of money finding out they don't work, or don't work well enough to meet inflated expectations. Sutherland showed sense in not continuing to build headsets, a lesson Sega and others learned the hard way.

Billions are spent on hardware that no one is even sure anyone wants, even if headsets no longer make us sick. Companies can't stop trying to push the virtual, digital world in front of our eyes in more invasive ways. Maybe what we want is to be able to tuck it all away in our pockets.

CHAPTER FIVE

Cyborgs And Brain-Computer Interfaces

While working for a business technology website in 2010, I went to a tech show called CeBIT in Hanover, Germany – that acronym stands for *Centrum für Büroautomation, Informationstechnologie und Telekommunikation*, or Center for Office Automation, Information Technology and Telecommunication. Given the excitement that generates, perhaps you'll not be surprised to hear that when CeBIT finally died in 2018, journalists hailed it as the end of the 'most hated tech trade show ever'.*

Such tech shows are held in bewilderingly massive halls. Each spring, CeBIT colonised the Hanover conference centre, the largest exhibition space in the world. That is not a good thing. It means walking 20 minutes to find the press room for lunch, only to discover there's no food left, before trekking 20 minutes in another direction for a meeting held in a booth full of scantily clad models left unbriefed about the vapourware product in their manicured hands. That year, desperate for a compelling, traffic-drawing story,

---

* All tech journalists will have their own opinions on what is the worst tech show. Personally, I hated them all because I worked for small teams and am naturally quite earnest, so did 15-hour days, churning out news obtained while marching up and down conference halls until my feet went numb, whether it was CES, MWC or IFA. That said, the upside of going to CES or MWC or IFA is that you're in Vegas, Barcelona or Berlin; with CeBIT you're in Hanover, and 'the only worthwhile thing to do in Hanover is leave', as my colleague claims to have told a cab driver, who politely didn't kick him out of the car. Thank you for buying this book so I don't have to return to that life. That headline is from The Next Web: https://biturl.top/JBNNby

I trudged up and down the aisles looking for anything genuinely newsworthy. I remember just one booth, but it was good: brain-reading technology.

Forget keyboards and mice and swiping with our fingers: imagine if we could just think, and our ideas would materialise in the digital world. That's what Austrian firm g.tec was demoing that year: its intendiX brain–computer interface (BCI) was a skullcap covered in electrodes that connected wirelessly to a laptop that interpreted the electroencephalogram (EEG) signals of brain activity into letters, to help locked-in people communicate. My colleague from a sister magazine was strapped into the system, trained for 10 minutes and then told to think of a phrase, focusing on letters one at a time. Amid the din of the crowded CeBIT halls, the system took 30 seconds to parse the signal revealing each letter, requiring five minutes to spell out on the display: 'howdy chums'*.

Nearly 10 years after that demo, Facebook's Mark Zuckerberg revealed the company was working on technology to let users think messages rather than type them. Elon Musk's Neuralink inserted implants into animal brains ahead of planned trials on humans after showing off a macaque monkey playing Pong with only the power of its mind to control its virtual paddle.

That's the extreme of augmentation, a long-term goal to blend machine intelligence with human brains. I'd rather keep my brain unadulterated and just type 'howdy' to my chums, but to each their own. But there are better uses for BCI. As with g.tec's project, much of the work developing BCI implants is for medical research and helping people with disease or injuries. And those with locked-in syndrome and the like may be more willing to subject themselves to surgery in the hopes of winning some form of communication, limited though it may be.

* You can read my story about this here: https://biturl.top/YZbeqi

There's more to melding tech and people than the brain, however. Many already live augmented by technology in other ways. Some amputees and those born without limbs sport smart prosthetics. But not all cyborg tech is so useful: thousands of people have had an RFID chip – the basic technology that lets you wave a credit card to pay for something – inserted under their skin to avoid carrying a card in their pocket.

We're talking about a few different ideas here. BCIs generally work by reading brain signals to interpret them, either with external pads attached to the head or an internal implant surgically embedded in the brain. As well as locked-in syndrome, this can also potentially treat symptoms of diseases such as Parkinson's and dementia, and has even helped a paralysed man learn to walk again. This is what the developers of products like Neuralink hope to use to connect human thought to machine intelligence.

But computers can meld with humans in other ways, be it cochlear implants, pacemakers and smart prosthetics, all of which use the body's electrical impulses for control. In these examples, need offsets the risk of surgery: if you suddenly can't walk, it may be worth going under the knife to regain that mobility. And it's not just limbs that go missing: hardware can be implanted (or merely worn) to regain or replace senses – we'll meet one artist who couldn't see colours but can now hear them thanks to a prosthetic antenna permanently attached to his head.

While details matter, the ethics of such efforts are clear: when designed to treat disease and overcome physical disability, this work can help people lead their lives as they choose. Take a step beyond this, and we're talking about augmentation – not replacing a missing sense, but adding or improving one – which shifts us into the world of cyborgs. So far, the most common additions are the aforementioned RFID chips placed under the skin to unlock doors or pay at the till without reaching into your pocket, though there's also sense-hacking, such as using a vibrating implant to train your brain to know when you're facing north. And some, including Elon Musk and the British professor who's deemed himself the world's first cyborg, believe that linking our bodies and technology will help us meat bags maintain superiority as AI outsmarts humanity. To them, it's not just our future. It's our only hope for a future.

★ ★ ★

But first, a quick jaunt into the past to see where this all began. The phrase 'cyborg' was coined by Manfred Clynes and Nathan Kline – and yes, Clynes and Kline do sound like they should be in a buddy cop film – in a 1960 paper called 'Cyborgs and Space' that argued it'd be easier to adjust humans to live in space than it would be to create 'an earthly environment for [them] in space'. The Rockland State researchers noted that animals could successfully alter themselves to meet the environment, though their biology limited them, using fish as their example: a 'particularly intelligent and resourceful fish' could escape water for land by developing technology to breathe — that's exactly what humans are doing when they travel to space.

That's quite the fish. To help humans survive space like our metaphorical fish, the researchers suggested integrated automated breathing devices built into the body. Such merging of machine and spaceman needed fresh jargon, and the term they proposed was 'cyborg', a portmanteau of 'cybernetic organism'. The first cyborg under this model wasn't a human but a rat with a pump installed under its skin that allowed chemical injections into it without bothering the rodent. Despite those otherworldly origins, 'cyborg' quickly began to mean something else, referring to any melding between human and machine.*

---

* 'I expected the word cyborg to survive,' Clynes said later in his life, with an *Atlantic* reporter noting the researcher was aware the word no longer meant what he'd intended. 'It's interesting ... to see how a word can have a life of its own.' Read more here: https://biturl.top/i6BzY3

There's an argument that any tool that expands the function of a human could make them a cyborg, but that means anyone holding a smartphone, wearing glasses or, heck, even shoes, could be considered a cyborg. Instead, let's restrain our definition to include humans with smart technology implanted or otherwise meshed with the body, be it an RFID payment chip, nerve-controlled prosthetic or brain implant. People sporting such technology may not personally identify as cyborgs, but the technology that assists them is the precursor to what people will use to become cyborgs, if anyone decides to take that route.

And by that definition, the first cyborg technology appeared just one year after that Clynes and Kline paper – not enabling humans to breathe in space, but helping a patient hear again. Not that anyone thought it would work at the time. The first implantable cochlear devices were created by William House, a physician in California. Cochlear implants are very different from a simple hearing aid, and can enable some hearing, though not necessarily all. In the second year of his practice, a patient of House brought him a newspaper clipping detailing the work of two French physicians, Dr André Djourno and Dr Charles Eyriès, who installed a coil of wire into the cochlear nerve of a patient to help him hear sound.

The patient, described in newspaper reports as a 'successful businessman', had become 'stone deaf', and had stayed that way despite two surgeries. He turned to Eyriès for help, and the doctor recalled from a few years prior the work of Djourno, who had meddled with the nerves in a rabbit so that he could make it move a paw when he played a sound. The pair teamed up and applied the principle to the patient, implanting an induction coil into the muscles on the side of the face and connecting it to the aural nerve. 'The surgeons then wired a low-frequency generator, which would transform sound waves into electric impulses of the same frequency, to an induction coil placed outside the man's temple, opposite the coil buried in his head,' one newspaper report said. 'Thus an electromagnetic field was set up between the two coils.'

It worked. The patient could hear sound again and was learning to differentiate words, though full spoken sentences were too much for the limited technology. He spent what reports described as 'the happiest afternoon for years sitting in the clinic listening to doors being slammed'. The implant eventually failed, and though the French researchers initially continued their work, Djourno eventually lost interest.

Inspired by the French example, House set to work building a cochlear implant, teaming up with neurosurgeon Dr Jack Doyle, who conveniently had a brother, Jim Doyle, an electrical engineer. The general idea behind such devices is that they consist of two parts. Externally, a microphone picks up sounds and converts them into digital signals that are sent via a transmitter coil to an internal receiver implanted under the skin, which turns the digital signal into an electrical one. That runs down a wire to electrodes implanted further in the ear, in a bit called the cochlea, where nerves take the signal and send it to the brain, where it's 'heard'.

First, the trio created a version to test on people who had become deaf after learning to speak, so they could tell House what sounds they heard; that seemed to work, so they set to developing a fuller device, which included an implantable electrode with a coil and amplifiers. The first two patients were implanted in 1961, and the system worked: they couldn't hear sounds clearly, but it was hoped that what they could hear would aid lip-reading.

The team ran out of cash and split apart.* Meanwhile House returned to his practice, but kept thinking about the implants. In 1968, he approached colleagues and ear-health leaders for help – but, one by one, each turned him down, saying the controversial work would damage their reputations, was

* House wrote in his memoir that he doesn't regret not filing for a patent on cochlear implants, but he'd be a wealthier man if he'd done so.

impossible or too expensive, or would further damage hearing. House had the sense, he wrote in his memoir, that his colleagues saw the implants as a cruel experiment on desperate people designed only to make money.

Indeed, cochlear impacts are still controversial, as the implants only return limited sounds, require training to work at all and are seen by some as a reflection of ableist assumptions that deaf people need to hear to fit into society. At the time, and even now, there are preferences regarding lip-reading, sign language and hearing aids over cochlear implants.

Controversy was familiar to House. He'd disrupted the industry almost from the beginning of his career by developing a microscope for ear surgery rather than just eyeballing it like his colleagues. His early attempts at surgery to return hearing to patients failed, and like the good researcher he was, he presented the negative results anyway, leading to mockery. His surgical solution for Ménière's disease was deemed unnecessary and for profit, while his plan to remove an acoustic tumour from one patient was met with an actual showdown with the staff neurosurgeon at the hospital, who resigned. In the end, the surgery was a success.

A year on, House had made progress, finding a collaborator in medical device engineer Jack Urban, who had helped build his operating microscope. The pair worked for a year developing a device: a battery-powered, five-electrode array, 25mm (1in) long and covered with silicone. The electrodes were set at 4mm (⅛in) intervals in the ear and met up in a plug that screwed into the bone behind the ear. They could control the sound frequency, intensity and timing.

As before, House ran tests on three adults who had become totally deaf after learning to speak, so they could more easily communicate to House what they heard. House and Urban tried different types of stimuli, and would then tweak the system based on patient feedback, trying different signal types and building new parts, before iterating again. House had to warn patients that no one had ever put an electric current this close to the brain for this long, so there was some risk.

Eventually, House and Urban settled on an amplitude-modulated (AM) signal – the same technique used by radios with that designation – and sent one patient out into the world with the cochlear implant for 16 hours. Despite his belief in the system, House worried that the critics would be right and the patient's hearing would be destroyed – but it worked.

The pair published their successful results, but were still met with hostility from the academic community, exacerbated by Urban's insurance company pulling support. Salvation came from House's daughter Karen. A documentary filmmaker, she filmed some of the test sessions, including one congenitally deaf 19-year-old who cried when she heard sound for the first time. House believes her film was the turning point for cochlear implants, with clips eventually shown on television broadcasts.

Though the debate continued – and still does – by 2009, 188,000 people had cochlear implants, and they are seen as one of many solutions for people with hearing issues. However, as noted previously, they remain controversial – partially because of scenes like the one that Karen House filmed. Video clips on social media show delighted toddlers hearing for the first time but don't reveal the limitations of the implants, causing a misunderstanding about what they can actually achieve for those with no or limited hearing. They don't work well for everyone and require training and practice to work at all. But having the option, poorly understood as this tool has always been, is better than not. And it was the first step to melding electrical pulses from the body with machines – making it the first real cyborg technology.

★ ★ ★

From hearing to the heart. The next machine-to-human augmentation was pacemakers, and keeping hearts ticking was unquestionably less controversial than hearing, with hordes of inventors, engineers and physicians racing to improve and perfect these devices.

Pacemakers have a simple but essential job: regulating the beating of a heart, ensuring it contracts to pump blood through the bodies of people for whom this organ doesn't beat at the correct rate or who have a blockage. Modern pacemakers do this by sending an electrical pulse on demand when the system detects no electrical activity in the heart's chamber, or using more complicated triggers depending on the cardiac problem. The pacemaker is implanted in the chest or shoulder and connected to leads that run into the heart, with a generator placed under the collar bone; when batteries need replacing, perhaps every 10 years, only the generator must be removed.

Experiments in using electrical pulses to contract the heart go back to before 1889, when John Alexander MacWilliam published his experiments in the *British Medical Journal.* He suggested that large sponges should be used as electrodes, the skin should be well moistened with salt solution, and that in medical situations where the procedure is deemed necessary, artificial respiration should 'by no means be neglected'. Good advice that's still true today.

In 1926 in Sydney, Australia, physician Mark Lidwill revived an apparently stillborn baby using a cobbled-together pacemaker plugged into a lighting point, with the electricity introduced via a needle electrode inserted directly into the heart. Flinch at that image, but the baby survived.

And in 1932, Albert Hyman coined the now-famous term with his own hand-cranked version, dubbing it the 'artificial pacemaker'. Looking a bit like a sewing machine, turning a crank generated electricity; like Lidwill's design, which he credited, the electricity was introduced to the heart via a needle shoved into the chest wall, so let's be glad that wasn't settled on as the best solution.

But then along came a series of innovations: in 1950, Canadian engineer John Hopps worked with doctors at

Toronto General Hospital to develop a wall socket-powered version using vacuum tubes, though computing developments soon allowed those to be replaced by transistors. American Paul Zoll is known as the 'father of modern cardiac therapy' for his work on external pacemakers. Though painful to use, by the mid-1950s, wires were placed in patients' chests to reduce the voltage needed. Colombians Alberto Vejarano Laverde and Jorge Reynolds Pombo experimented with battery-powered devices – the latter using, at one point, a car battery – as connecting to the mains had some clear downsides if the power went out. In 1958 American Earl Bakken built the first wearable pacemaker, helped by the arrival of the silicon transistor for controls. A few years later, British physicians developed a similar version worn on a belt with a rechargeable battery.

All of these were worn outside the body – well, in the case of Lidwill's needle and the odd wire, mostly outside – so none of the recipients qualifies as cyborgs, a fact that probably bothered none of them. But in 1958, the first internal pacemaker was implanted at the Karolinska Institute in Sweden. Designed by engineer Rune Elmqvist and surgeon Åke Senning, it failed after eight hours. Don't worry about the patient though: Arne Larsson outlived both Elmqvist and Senning and had 26 different pacemakers throughout his life – and if that doesn't qualify as dedication to cyborg life, nothing does.

What would make someone take the risk of becoming the first person with an implantable device controlling their heart? Larsson had suffered a viral infection that scarred his heart, leaving it beating just 28 times a minute. The decreased blood flow often caused him to faint, and his wife had to resuscitate him with manual compressions up to 30 times daily. So even though Elmqvist and Senning didn't think their pacemaker was ready – and arguably it wasn't – she urged them to try. After the first failed, they implanted their only backup, which worked for three years.

Those pacemakers didn't immediately lead to a commercial product, partly because we needed technology to catch up, with silicon transistors making the required tiny battery-powered circuits possible. Plus, we needed Wilson Greatbatch. He spent the Second World War in the US Navy, including flying combat missions. Returning home unsure what to do, Greatbatch got a telephone repair job before taking advantage of the GI Bill to attend university at a discount in exchange for his service. He headed to Cornell University, eventually working on the animal behaviour farm, fitting sheep and goats with devices to measure blood pressure, heart rate and brain waves for experiments. In 1951, he had lunch with visiting surgeons who discussed heart block, when the upper chambers of the heart can't communicate to the lower chambers, which causes arrhythmia and can be fatal.

Greatbatch saw that as a communications problem akin to fixing radios in the Navy and believed the solution could be neater than the 'big TV-sized boxes that were plugged into the wall with the end of the wires running over to the patient,' he said in an IEEE oral history interview.* 'That patient's world was the length of that extension cord.' There were already alternatives to those heavy units, but Greatbatch saw failings with the wearable pacemakers, too. Made by Medtronic, a company founded by the aforementioned Earl Bakken, these versions were worn on a belt with wires that went through the skin, meaning patients couldn't swim, take more than a sponge bath or sleep on their fronts – better than being attached to a big box connected to a wall, but still limiting. 'We engineers have not to this day learned how to

---

* This is from an interview with Greatbatch conducted by Frederik Nebeker at the IEEE History Center in the US: https://biturl.top/M3q6nm The IEEE History Center has a collection of more than 800 oral histories in electrical and computer technology here: https://biturl.top/eeyeMr

run a wire through the skin and have it seal,' Greatbatch said. 'It's always an open wound.'

Greatbatch spent a decade working on his own implantable pacemaker designs with two colleagues, William Chardack and Andrew Gage, with assistance on materials from Medtronic, which eventually licensed the design. While working in Buffalo, NY, Greatbatch grabbed the wrong resistor for a circuit designed to record heart sounds, realising it could work for pacemakers. A few weeks later, in 1958, he tested it on a dog.

In 1960, the team published details of the first successful implants in *Surgery*, sharing everything they learned in medical journals, at research seminars and even sales visits to hospitals. Within five years, implantable pacemakers became the default way to treat blocked hearts – a pace that surprised even Greatbatch, who noted, 'that was almost unheard of in the industry'.

For the first 10 years, Medtronic manufactured the pacemakers, with all designs overseen by Greatbatch. But the company split over batteries. The first models were powered by mercury batteries bought from a company called Mallory. Mercury batteries need to release gas to keep the reaction balanced; that could be achieved by using a permeable casing, but that let *in* water vapour, which wasn't ideal and limited a pacemaker's lifespan to just two years. As patients lived for an average of six more years, they faced three pacemaker surgeries; Greatbatch realised that improving battery life to 10 years would have meant no replacements were required for most patients.

Medtronic wanted to keep using the Mallory batteries, so bought the necessary patents and continued down that road, while Greatbatch's team considered creating batteries that could be recharged through the skin, as well as nuclear batteries using plutonium. They worked a treat, but as that's a controlled material they would have needed to track every pacemaker-implanted patient as if they were a nuclear weapon.

The solution came in a phone call from the Catalyst Research Corporation, which had been making lithium-and-salt-based batteries for the military, but believed the idea could have further uses. Initially, Catalyst was going to build the batteries, but they couldn't hermetically seal them, and some were destroyed in the mail to Greatbatch's new, eponymous company. Catalyst decided to set up a new factory to make and seal the batteries, but it was located near a stream and flooded. 'When the water hit the lithium and sodium, the whole factory blew sky high,' Greatbatch said.

So his own company started production and at its peak made or licensed almost all of the batteries that went into pacemakers. One of the first using a lithium battery was implanted in an Australian, who then wandered into the outback and disappeared. Twenty-two years later, they found him alive – and his pacemaker had been running just fine the whole time.

Now, a million pacemakers are implanted each year around the world and many more would be if they were cheaper to make or could be reused. More recent innovations allow them to be controlled or at least monitored by smartphone apps. On the downside, the US FDA has warned that smartphones can disrupt pacemakers, so hold your iPhone far away from your chest when you adjust your settings in your future smart pacemaker.*

★ ★ ★

Prosthetics have been used for thousands of years, with archaeologists digging up a false eye made from bitumen and gold from 3000 BC in Iran, and false toes and feet from 1000 BC in Egypt. Pliny the Elder wrote in AD 77 about Roman

---

* Greatbatch's barn that he worked in has been lovingly restored and maintained by the Historical Society of the town of Clarence, in the state of New York.

general Marcus Sergius being able to retake a battlefield circa 200 BC with an iron hand. By 1505, little had changed, so when Götz von Berlichingen lost a hand in the Bavarian battle known as the Siege of Landshut, he too was given an iron hand, though this heavy addition was attached to his armour rather than his stump, and had flexible joints that let him take the reins of his horse. Like his Roman predecessor, the German used the false limb to head back into battle.

Around the same time over in France, Ambroise Paré designed mechanical hands with springs and catches that – once again – allowed an army captain to return to battle. Prosthetic hands were made for less violent reasons, too: in 1600, Italian surgeon Giovanni Tommaso Minadoi tells of a prosthetic hand that allowed the wearer to write with a quill and untie their purse. Generally, prosthetics were rare for two reasons: they were expensive; and most people who lost limbs died for lack of effective medical treatment.

The development of bionic limbs began with the shift from unpowered models towards externally powered designs. These became more common in the 1950s and 60s, with research driven by an increase in amputees following the Second World War as well as the thalidomide scandal leading to the birth of more than 10,000 children with deformities that included missing limbs. In Edinburgh, researchers developed pneumatic limbs, making it easier for children to control their prosthetics.

A few inventors were ahead of their time. In 1948, Reinhold Reiter, a student at Munich University, designed and built the first myoelectric prosthetic limb, which used nerve signals from the arm to control the device, though no one appears to have paid much attention at the time, despite him effectively inventing wearable robotic arms. And in the 1950s, IBM engineer and automotive crash test dummy inventor Samuel W. Alderson developed a battery-powered arm that used small motors based on the ones he'd helped develop for missile guidance systems during the war. Switches built into the sole of a shoe controlled the electric arm, with 12 possible signals. Full pressure from the big toe would flex the elbow, while light pressure on the little toe turned on the electric motor. It was another 12 years before Russian scientist Alexander Kobrinski produced a myoelectric hand, helped by the arrival of transistors for control. Though the 'Russian Hand' offloaded the electronics and batteries on to a belt, it was still heavy, expensive and suffered signal delay. In 1998, David Gow, the head of the Bioengineering Centre of the Princess Margaret Rose Hospital in Edinburgh, fitted patient Campbell Aird with a full arm replacement, including the first electric shoulder. He predicted it would cost about £10,000 commercially and hoped that the shoulder and the follow-up myoelectric bionic hand would one day be available on Britain's National Health Service (NHS), which finally happened in 2022.

At this point, prosthetic limbs were devices to wear, with sensors in the device picking up signals from remaining muscles via electrodes. The shift to technologies embedded in the body came with neural interfaces, in which electrodes are implanted into muscles or nerves to pick up electrical signals that are understood by the prosthetic, letting the user (once trained) simply think to move their hand or arm or foot.

The first to benefit from this idea was Jesse Sullivan. In 2001, this electrician was at work fixing power lines for the Tennessee Power Company and made a mistake, touching a live cable. The shock of 7,000 volts of electricity led to both his arms requiring amputation at the shoulder. Initially, he was given traditional prosthetic arms. 'When they first gave me a prosthesis it was Second World War technology and I was devastated,' he told the BBC. Shortly after, Sullivan was given a much smarter solution: for his left arm, the very first bionic prosthetic embedded into the body, from the Rehabilitation Institute of Chicago, then led by Dr Todd Kuiken. Initially, the robotic limb was controlled via neural signals from his shoulder, but sensitivity there meant the sensors were shifted to his chest, where four nerves that led to his arms were grafted into place, letting him move the limb, feel how hard his hand is squeezing and feel temperatures. He can now even tie his own shoelaces.

To close his hand all he has to do is think about it, and signals sent from his nerves to the prosthetic do the work. The technique, called muscle reinnervation, means nerves that once led to Sullivan's arms are grafted to his pectoral muscles, where that thought to close his hand is picked up by electrodes that deliver the signal to the computer system on the bionic arm. It took months of training after surgery for Sullivan to learn how to move the hand well enough, but years later he said he no longer needed to consciously even think to make it work.

When he was given his first bionic arm, Sullivan was told to give it a real-world test by the researchers who built it: 'Don't bring it back looking new.' He took that to heart, breaking one arm when he was starting a lawnmower. The research team led by Kuiken added DARPA to its cadre of partners, in part because of the number of military amputees returning from Iraq, and the US military later expanded its work as part of a project called Revolutionising Prosthetics, pushing the industry forward for the most terrible of reasons.

The first woman to receive a bionic arm was Claudia Mitchell, who lost her left arm at the shoulder in a motorcycle crash in 2004. At the time, her first bionic arm, an updated version of Sullivan's, cost $60,000. Not only could she use the prosthetic to pick up objects and so on, but she could also feel sensations such as heat. That required nerves from her arms and chest to be connected, and this time that included nerves that feel sensation in the skin, so that when that patch of skin is touched, she feels it as though it's the skin of her missing hand. Targeted reinnervation can have a faster response time than myoelectrics and allows multiple joints to move simultaneously, though training is required.

Prosthetics have taken further steps forward. In 2016, I met a series of amputees at an event talking about 'prosthetic envy' – techno-fetishism, where people with all their limbs desire to 'upgrade' the limbs they were born with in favour of gadget-filled, carbon-fibre models. While the latest bionic limbs may offer abilities that flesh doesn't, they're still not as impressive as many imagine, noted James Young. In 2012, Young lost an arm and a leg in a train accident. Four years later, he had a new prosthetic arm, inspired by the game *Metal Gear Solid* and its main character Snake's bionic arm. Unquestionably an odd marketing gamble, the project was paid for by the game's maker Konami.

Young's arm features a USB port so he can charge his phone, a small display to check messages and a built-in laser pointer. The design is slick and futuristic, with LED lighting running down its length, and Young told me people would run up to him and excitedly proclaim him the Winter Soldier from the Marvel films. But it looks cooler than it is, he added – and despite the tech, is more limited than a human arm.

Another of the speakers at the event was Nigel Ackland, and he shrugged off the suggestion he's in any way enhanced or a 'cyborg' because of his robotic hand, the Bebionic hand made by Ottobock. The hand does have abilities not found on a flesh-and-bone limb, including the ability to spin 360 degrees, which he finds convenient for fetching mugs from a funny angle in his kitchen cupboard. But there are limitations. 'I can't feel my wife's face,' he told me.

Such bionics are remarkable leaps forward for those with missing limbs, but it's worth noting that their cost remains high and that means most people aren't getting smart prosthetics, but basic models – 'a hook, controlled by a bit of string and a rubber band', Ackland said. Wonderfully, organisations such as Open Bionics are stepping in to help. This UK-based company is working to offer smart bionic arms at a lower cost using 3D printing – and, delightfully, has royalty-free licensing deals with Disney for superhero-themed designs, meaning children lacking limbs get to look as amazing as they are. Indeed, sometimes technology is less about function than form.

Cochlear implants, pacemakers and prosthetics are all augmentations to help people recover an ability, but their development has also moved forward the technologies required – be it surgical implants, medical-grade materials or electrically manipulating the body's nervous system – to create cyborgs. Let's meet some.

★ ★ ★

British professor of cybernetics Kevin Warwick considers himself the first cyborg, despite admitting that someone with a smart pacemaker or cochlear implant could make a solid claim for that self-bestowed title, not least as they retain their implants. His own implant was removed just three months after it was surgically embedded in the median nerve of his arm and used to turn off lights, control robots via the internet and even communicate with his wife's implant. Warwick may not be a cyborg, but he's clearly a romantic.

Warwick landed his first professorship in 1988 at the Cybernetics department at the University of Reading, on the promise he'd make the department an international player. In the UK, funding follows students, so Warwick knew he needed to increase the department's enrolment while simultaneously attracting government and industry investment for ongoing research to keep staff paid. A sexy, attention-getting project was the answer. Warwick had previously developed robots, including a controllable hand and a self-organising group of tiny mobile bots called the Seven Dwarves.* But he'd also

* For some of this, Warwick worked with a researcher called Mike Brady, previously of MIT in the US, saying they spent some time looking into mobile robots in military uses, in particular as part of DARPA — potentially referencing the Shakey research we explored earlier.

worked with assistive technologies, developing an auto-adjusting walking frame that lessened its support as people became stronger; a phone system for deaf people; and glasses that people with epilepsy could wear in the bath that would send an alert to drain the water if they had a seizure, to prevent drownings. He then built a robotic chair for Jimmy Savile, the television presenter who worked with the National Spinal Injuries Unit at Stoke Mandeville Hospital.*

Instead of building on all of that work, Warwick decided to experiment on himself and become the self-proclaimed world's first cyborg. Why? The technology was ready to go – it was mostly for medical or research purposes, but it could be done. And he thinks that means you should make it happen. 'That's what science and research is all about,' he tells me. 'As a scientist, you are a science fiction writer because it's not been done until you do it.'

The first implant Warwick tried was a simple RFID transmitter, just 2cm (¾in) long and housed in glass – he shattered the first one trying to sterilise it through boiling. 'I dropped one on the floor, and it was a rubber floor, not a hard floor, and it just broke and shattered and went all over the place,' he tells me. 'In hindsight, what if this happened inside my own body?'

In 1998, Warwick's own GP embedded the device in his arm below the skin and fat, right along the muscle; it took two tries to find the best spot. Of course, this isn't the normal service a GP provides. Warwick tells me the doctor received two letters from medical authorities, one after the other. 'One was saying you really shouldn't have done this,' he says. 'The other was saying congratulations, well done.'

---

* This was before public claims about Savile's horrific criminal behaviour came to light – though Warwick does show some insight into Savile's character, at least, describing in his memoir the child abuser as someone who knows how to manipulate the media.

Once there, the RFID transmitter meant Warwick's approach could open a door and trigger it to say a welcoming message, turn on a light and load his webpage on his computer. None of that will be impressive to modern audiences, who can do this all via their voice or phones, but at the time it was certainly a neat trick. However, none of these showpieces worked four days before the press launch. A bug meant the chosen code emitted from the RFID transmitter – 666, owing to a lab technician's devilish sense of humour – was failing to pick up. But other codes did, spookily enough, and after hard graft by the team, everything worked by the time the media arrived for Warwick's performance.

The RFID transmitter was basic – he could have simply carried it in his hand to achieve the same effect – and that experiment didn't scratch the itch for Warwick. Nine days later, the RFID tube was removed, and for his second go at being a cyborg, the team decided to install a mini array of microelectrodes into the median nerve of his wrist. Designed by Richard Normann at the University of Utah and then made by Bionic Tech, the 'Utah Array', which was later rebranded as 'BrainGate', had 100 silicon spikes to mesh with nerves to capture electrical signals, sending them up wires that exited the body to connect to digital reading equipment. If it worked, Warwick would move his finger, with the array capturing and sharing the resultant signal. It could later be played back, hopefully moving the same finger. And if that worked, they could start to meddle with the signal to make other things happen – they even considered getting a few drinks into Warwick to see if they could electrically simulate being tipsy.

There was a problem: until that point, the array had only been used on animals, predominantly cats, which were subsequently killed to study the impact on their nervous systems. Not only was it unclear what side effects Warwick faced, but the Reading team wasn't sure if the manufacturers would even sell them an array for human use, so they didn't mention it. (They eventually spotted it in the media, and were generally pleased, Warwick said.)

There were further problems – Warwick said he felt like everyone involved, even tangentially, was dragging their feet on his experiment, with challenges from ethics committees, funding drying up and slow delivery times all slowing progress. But eventually, despite such lack of enthusiasm from everyone except the man being cut up, the surgery was arranged. To install the array, which sort of looks like a high-tech hair brush, a 5cm (2in) incision was made below his wrist to look for the median nerve, which controls movement in the hand. Into that cut, the surgeon shoved a tube further up into Warwick's arm through which they pushed the 3mm by 3mm (⅛in by ⅛in) array and its trailing wires, which were in turn fixed to a pad of connectors that sat outside the body, leaving via a hole in his skin that would stay open for the duration of the experiment.

The procedure is an unpleasant reminder of the blood and gore that makes up our bodies; machines are the neat and tidy side of cyborgs, not us. During the surgery, the surgeon bumped the median nerve while meddling with a blood vessel that was in the way, sending a shock down Warwick's arm; he yelled out in pain and euphoria, telling the assembled BBC reporters that it highlights how humans are driven by electricity, just like robots.

Then, the tiny array got stuck in the tube embedded in his arm. Pushing didn't budge it. Nor did suction. Finally, they washed it out with water and got the hardware into position. All that was left now, before stitching the professor up, was to install and connect the array. This was done by whacking it with a special machine called an impactor unit so it slammed into his nerve with enough force to embed 100 bristles.

It was unclear whether the array was working until Warwick returned to the lab. The lab's self-designed interface went bonkers, with 'a multitude of small electrical currents scurrying down [his] arm from the terminal pad in the direction of the array'. LEDs on the hardware were suspected (accurately) as the culprit, so the team switched to the interface supplied by Bionic Tech; it also went bonkers. This time, it wasn't tiny lights scuppering the work but an engineer's mobile phone receiving a text message. Lights removed and phones disabled, initial tests suggested the surgery had been successful enough, as there were 20 working channels out of a potential 100 – sufficient for the planned experiments, once issues were addressed around frequency and noise reduction.

So what can you do with networking equipment shoved into your nervous system? The system was mapped to Warwick's hand movements, though it didn't track his fingers but the electrical signals sent from his brain to his hand to tell his fingers what to do. That could then be used to control external devices. At rest, the computer showed a red light; moving fingers turned the screen green. Mapped to sound, when Warwick moved his finger back and forth, the computer produced a sound like a steam engine; moving faster, it would chug faster. For media, the Reading lab assembled a host of demos: controlling an artificial hand via Warwick's own (he would clench his fist; the robot hand would follow suit); switching on a lamp, coffee maker and alarm that were network-connected; and driving a Lego robot. He says in his memoir that it was like using a TV remote control, or to be more accurate, like being a remote control.

The assembled newspaper media were suitably impressed but criticism had started to pick up in specialist publications, including the excellent, cynical British tech website The Register, which called Warwick 'Captain Cyborg' and a 'media strumpet'. And while it's true the project was limited – especially the RFID chip – it's also true that no one had done it before. A stunt can also be science. Warwick admits that he wasn't doing the bulk of the technical work. Mark Gasson developed the challenging electronics, the hardware was bought in and others sorted the robotics. Warwick was the convener, perhaps, and the test subject, of course – but he also saw himself as the communicator with the 'outside world'. And getting science stories picked up by the media is no easy task without a bit of hype to nab headlines. He tells me: 'It was easy for some people to grumble and say I'm just doing it for publicity, which well, I wasn't doing it *just* for publicity, but I wasn't doing it to not have publicity. That was an important aspect.' Either way, he says he achieved his goal: more students signed up for his department.

Undaunted by the media criticism, Warwick followed with more demos – also with press involvement – showing him controlling the Reading lab's robot hand from New York and changing the colour of a smart necklace his wife wore with a squeeze of his hand. All of that was one-way, however. Warwick wanted to see if it was possible to stimulate his hand to move or react by feeding current into the implant. Previous attempts to do this had, perhaps sensibly, been on animals, so it was initially unclear how much current to use. They started low, using half as much as a cat's sciatica nerve required – 10 microamps for Kevin, 20 for the poor kitty – and gave Warwick a dead man's switch in his right hand in case it hurt. Inexplicably, they constructed the switch out of Lego.

At 10 microamps, nothing happened. They turned it up, and still nothing. After new components were installed, he finally felt a tightening of muscles in one finger at 80 microamps – it was hardly the response expected, and suggested the plans to simulate emotion or drunkenness weren't likely to be possible, though they did successfully feed ultrasonic sound in so he could 'see' large objects when blindfolded. Indeed, the finger reaction itself only had an 80 per cent success rate. After testing the electrodes, the team realised just seven were still working. They cleaned the contact pad, hoping that would help the success rate, and that killed another two. Whoops.

With only a few electrode bristles left functioning in Warwick's arm, the race was on to attempt the final experiment. The hope had been that Warwick's wife Irena could have the same style implant as him, but ethics committees and the practicality of installing one and getting it up and running in time put paid to that. Instead, Irena had a pair of needles jabbed through her wrist into her median nerve in order to pick up signals. This allowed the pair to communicate via their nervous systems – well, it let Irena twitch her fingers, which the system translated to pulses to send to Warwick's implant. 'It worked, we were able to communicate – it was like Samuel Morse doing the first telegraph message but nervous system to nervous system,' he tells me.

The surgeon extracted Irena's electrode needles the same day, and soon removed the implant from Warwick's wrist. That was in 2002, and more than two decades later, Warwick has never had another piece of technology surgically implanted into his body. He's surprised no one else has followed up his work. 'I'm completely disappointed; I felt I'd gone out on a limb,' he says. 'So for me to go ahead with another implant at that time, or even five years later, I'd have been doing it almost for my own sake, my own interests.' Instead, his work has shifted to the use of implants to address diseases like Parkinson's, first for monitoring but eventually, perhaps, for treatment.

Thousands of others have followed Warwick's lead, not using electrodes but RFID chips, having them embedded under their skin to tap to pay at the till or open doors. So far, with a few notable exceptions of medical wonders and cutting-edge artists, that's as far as mainstream cyborg life has extended. That disappoints Warwick – and it means his prediction of a world divided into cyborgs and luddite humans by 2050 is likely far from the mark.

* * *

There are a few people who have built smart machines into their own bodies to add rather than just replace. One of those people is Neil Harbisson, an artist based in Northern Ireland and Spain. While Warwick's implants were simple and embedded, as much as possible, inside the professor's body, Harbisson's hangs over his head like the prey-attracting light of a deep-sea anglerfish, an effect furthered by his bowl haircut with undercut.

Harbisson was born colour-blind. After seeing a lecture on cybernetics given by Adam Montandon, a design engineer at the University of Southern Denmark, while studying at Dartington College of Arts in England, Harbisson realised technology could give him the colours that the rest of us see, not visually but through sound. Harbisson approached Montandon and the pair collaborated, with the latter developing software that expressed colours in sound, capturing the scene to be translated with a camera worn on Harbisson's head that connected to a 5kg (11lb) computer he wore in a backpack. The sounds were played via headphones. 'At first, I had to memorise the sound of each colour, but after some time this information became subliminal, I didn't have to think about the notes, colour became a perception,' Harbisson writes in an essay.* 'And after some months, colour became a feeling, I started having favourite colours.'

That was only the beginning. Next, computer scientist Peter Kese added volume levels to represent saturation and expanded the software to include 360 hues, represented by notes. Rather than a head-mounted camera, Harbisson shifted to what he calls an antenna, an arcing design that holds the camera above his head, so he can see all around him. And he ditched the headphones for bone-conducting technology. By 2010, he no longer needed a backpack filled with a computer, as technology had miniaturised enough that a single chip attached to the back of his skull was sufficient to translate colours to sound and send them through his skull to be heard, which he's dubbed sonochromatism.

At this point, Harbisson had successfully added a sense to his body, but it wasn't in any way permanently attached – he

---

* This is Neil Harbisson's collection of essays: https://biturl.top/YNFrq2

wasn't a cyborg, he had a one-off, very impressive wearable computer. The next step was drilling into his skull to attach the antenna permanently. Finding a surgeon to do the work took two years and had to be done 'underground' after a bioethics committee refused the procedure.

The antenna is implanted into his upper occipital bone, which is at the back of the head. It took two months to heal post-surgery, but the senses he's gained for hearing colour are far beyond what the rest of us can see, such as ultraviolet and infrared. Now, he can paint what he hears. Cities, he reports, aren't grey at all – Madrid is amber terracotta, Lisbon is yellow–turquoise, and London is very golden red – and humans are shades of orange. Looking at a Picasso sparks music, but so does the food on his plate – he hopes to open a restaurant one day where people can have pop songs for starters, concertos as main dishes and perhaps a bit of Björk for dessert. He no longer dresses to look good but to sound good. 'If I'm happy I dress in C major,' he writes. 'If I'm sad I dress in a minor chord. So if I need to go to a funeral, I might dress in B minor (that's turquoise, purple and orange).' This works surprisingly well: he's a snappy, if bold, dresser.

Beyond colours, he can feel if someone touches the antenna, comparing it to someone tapping your teeth or nails. It must have rather hurt when he was at a protest and policemen tried to pull it off his head, believing him to be filming them rather than listening to colours.

All of this impacts Harbisson's art. He can paint the colour-sounds he hears, creating colour sets of objects, cities and people. He gives what he calls 'face concerts'. Essentially an instrument, Harbisson can hook himself up to loudspeakers and play the sounds made by his audiences' faces – and if it doesn't sound good, that's their fault, he notes.

Harbisson fully identifies as a cyborg. In 2004, the artist applied for a new British passport and was told he had to take the antenna off; after arguing it was a part of his body, the authorities – presumably not really sure how to proceed – carried on with issuing the document, complete with photo including the antenna peeking out over his hair.*

Harbisson bridges the world between disability and augmentation; he lacked the ability to see colours, but with his antenna, extended his vision well beyond what the rest of us can do.

For some, augmentation is a matter of life extension: Peter Scott-Morgan didn't want to extend his abilities – but he wasn't ready to die. So, in the years between a terminal diagnosis and death, he used his considerable engineering and robotics skills and subjected himself to a host of procedures that left him part human and part machine.

In 2016, Scott-Morgan noticed that he struggled to dry his feet after a shower. That small moment of trouble heralded a diagnosis of ALS, the same neurodegenerative disease Stephen Hawking had. Scott-Morgan was given two years to live – and worse, for much of that time he'd be unable to walk or talk. This was not a diagnosis that Scott-Morgan was willing to just accept, and he had the power to do something about it: he held the first PhD in robotics from a British university and had been working in the field for decades, arguing that one day humans would swap out failing parts with machines. Now, his own life – or at least, the quality of his remaining life – relied on that very belief.

In his memoir, *Peter 2.0: The Human Cyborg*, Scott-Morgan recalls meeting the specialist nurse, Tracy, who would help manage his care. She walks him through how the disease will affect his body, and what care he'll need. For example, he'll remain continent, but will lack mobility. Put another way, he can use the bathroom on his own, he just can't get there. That aside, two of the core symptoms of ALS are weak muscles

* To be clear, nowhere on the passport does the government call him a cyborg.

around the lungs preventing breathing, as well as difficulty eating and drinking.

He wrote in his memoir that he needed to be 'replumbed'. By that, he means surgically inserting a tube to take over his eating and drinking, as well as two 'output' tubes for urine and faeces. That would involve damaging healthy organs, but would give him more bodily control. Then there was a laryngectomy, allowing his breathing to be controlled by a machine but costing him his voice box, addressing one of the biggest challenges faced by those with ALS.

That removed his ability to talk, so a company called CereProc pre-recorded 20,000 words for him to be able to play back.* That voice synthesiser would be controlled by eye-tracking technology, so he could 'type' by looking at keys, with words and phrases suggested by AI that was listening to whoever was speaking to him, like a personal, medical Alexa.

When his facial muscles failed, he devised an avatar to be projected on to his own face; though his husband Francis suggested it be projected on to his chest instead. For that, the couple turned to a company called Optimize3D, based at the famous Pinewood Studios, where *Star Wars* was filmed, among others. Scott-Morgan's face was covered with dots for motion capture – they used Max Factor mascara – and then he was recorded by more than 50 high-definition cameras while speaking, as well as making every facial expression he could muster.

This was beyond survival. Forget staring at a boring nursing home ceiling, he wrote. Instead, he wanted to be able to live

* In one of the five-hour recording sessions, he writes in his memoir, Scott-Morgan was able to record a set of phrases of his own choosing. He opted for a list of synonyms, beginning with 'screw-up' and ending with 'cluster-fuck'. The documentary producer tagging along apologised, saying noise was picked up and the set would have to be re-recorded – the noise was his own laughing.

and explore cyberspace in virtual reality. Scott-Morgan's body would fail him, but his mind would not; he planned to compose a symphony, write a book and create art, all inside virtual reality – the only place he would be able to 'move' and where he and his husband could interact. He recalls in his memoir saying to Francis that most people use VR to escape their lives, but he wanted to reclaim theirs.

Scott-Morgan had further plans. He ordered a Permobil F5 Corpus, a cyborg harness and robotic life-improving exoskeleton, essentially a standing electric wheelchair with additions to make it operate more easily for someone with no mobility. At one point, he trialled using a robotic arm that his own fit inside, controlled by eye tracking. It took an hour to learn to pick up an orange and place it in a bowl, but by the end he was able to reach out to grip Francis's hand.

None of this came easily. Surgeons said no. Charitable organisations backed out of support, leading Scott-Morgan to set up his own. And all the while Scott-Morgan lost more of his physical abilities. His argument wasn't that technology could save anyone from ALS, which would eventually kill them in some way. Instead, he believed that people with the disease had the right to make the best of their lives using the tools available to them – they had the right to thrive. There were options, and he wanted people to know that. His road was hard and expensive. But it was possible.

Scott-Morgan passed away in 2022, gaining more years than doctors predicted and maintaining his speech and other abilities longer than expected. A few months before his death, he tweeted about how lucky he was and how much fun he was having. Scott-Morgan didn't get to live forever, not even as long as he wanted, but there's no better reason to merge man and machine than his.

Indeed, most people taking steps towards cyborg existence aren't doing anything anywhere near what Scott-Morgan accomplished. But along with those with temporary night vision thanks to chemicals found in deep-sea fish or the ability to 'hear' Wi-Fi, there are plenty of DIY cyborgs shoving magnets under their skin, installing USB drives in place of missing fingers, implanting RFID chips into fingers to hold payment cards and ticket data, and wearing belly-button piercings that vibrate when you face true north, for those easily lost. And then there's Stelios Arcadio, better known as Stelarc, the artist who has electrodes attached to his muscles to play music and who grew an extra ear on his arm, as you do.

* * *

But none of these small augmentations are going to stave off an AI apocalypse, now, are they? And that's what Zuckerberg and Musk and the rest are hoping to accomplish with their brain–computer interfaces – though plenty of serious researchers are merely looking to treat serious diseases.

Since the 1700s, we've understood that the body's electrical impulses can be read, and British researcher Richard Caton read electrical signals from living animals' brains using what he dubbed a galvanometer in 1875. The first time an electroencephalogram (EEG) was recorded was in 1912 by a Ukrainian physiologist Vladimir Pravdich-Neminsky, though that was again from animals; the first human recording was 12 years later, in Germany. But the specific idea of linking a brain and a computer via an interface – that B, C and I in BCI – comes via Jacques Vidal, who coined the term in 1973 while a professor at the University of California, Los Angeles, followed by implanting monkeys in 1987 and humans in 1998. His lab showed BCI can be used to control a cursor, with a research participant navigating a digital maze.

Now, there are dozens of stories of these technologies changing lives, often with the upgraded version of the Utah Array, the hairbrush-resembling set of sensors that Warwick used in his arm. About 6mm (¼in) in length on each side, it has 100 or more silicon electrodes, essentially tiny needles about 3mm (⅛in) long. Each needle is an independent electrode, meaning each can listen to a separate group of neurons. 'This device can listen selectively to the neural activity in hundreds of neurons with unprecedented selectivity,' Normann explains in an interview, adding that the aim with the array is to give neurologists treating patients with disorders or injuries of the nervous system a 'new set of tools to, really for the first time, allow them to actually treat these pathologies rather than just simply diagnosing them.'*

In 2004, Matt Nagle was implanted with a Utah Array, letting him change the TV channel and otherwise interact with electronics, controlling a bionic hand a year later. Dennis DeGray fell after slipping on mould on the way to take out the bins, causing a neck injury that paralysed him from the shoulders down. Ten years later, in 2016, Stanford University researchers embedded a Utah Array into his brain, giving him the ability to move a cursor on a computer screen and eventually to type. In 2021, researchers at the University of California, San Francisco used a neural implant to let a paralysed man speak at a rate of 15 words per minute.

None of this technology is new. The Utah Array is manufactured by Blackrock Neurotech, which has been chipping away at BCI for a decade longer than Neuralink. And its roots go back even further, as Blackrock Neurotech uses technologies hoovered up after an even older BCI company, Cyberkinetics, went out of business in 2000. Founded by Florian Solzbacher and Marcus Gerhardt, Blackrock Neurotech's NeuroPort BCI implant has been used to let paralysed people control a robot to feed themselves, regain control of their own arm and wrist, sense touch, and even control two arms at once with implants in each side of the brain. Such work is done alongside researchers at universities, with current efforts to build a device that works

* You can watch a video of the interview on YouTube here: https://biturl.top/3QNfm2

more easily at home described as a 'trial'. This is still very much science rather than engineering.

But BCIs clearly work. It's just a matter of perfecting the surgery and making them small enough, safe enough and cheap enough to help all the people with locked-in syndrome, those who are paralysed and those who have brain diseases – and, of course, understanding what actually helps such people.

In 2016, Elon Musk quietly established Neuralink with an aim to engineer a commercial product, hiring neuroscientists to set up a company to build a new way of doing BCI. The initial aim was to address diseases or brain injuries, but from the outset, Musk made the long-term goal clear: 'symbiosis with artificial intelligence' to avoid the unchecked advance of AI causing an existential threat to humans. He doesn't appear to be joking: as we saw earlier, Musk and others in tech are genuinely concerned that superintelligent AI will outsmart and supersede people. If we can embed AI capabilities into our brains using this sort of technology, the argument goes, we can keep up with machine intelligence. Personally, I'm not sure how this wouldn't just give those superintelligent, diabolical AI systems direct access to our brains to manipulate us for their own devices, like those parasitic wasps that control cockroaches.

Neuralink held two high-profile demonstrations – more akin to the launch of a new smartphone than a scientific discovery – showing a pig with a brain implant (spoiler alert: it looked like a pig) in 2020 and, in 2021, a monkey playing Pong using neural signals. Those demos were criticised for being akin to what scientists were doing 20 years prior; however, Neuralink isn't about proving the science but engineering a commercial device. Trials in humans were projected to begin in 2020, but that was delayed with the first implant in 2024.

Neuralink is working to develop neural lace, very fine electrodes that wouldn't require full open brain surgery to implant in a layer above the cortex. And that's core to what Neuralink is doing: it's about engineering systems that enable

BCI to step away from science and into a commercial product, with advancements in designs that mean less invasive surgery is required. Its rivals have the same aim, with one product, the Synchron Stentrode, being implanted into a blood vessel, negating the need for more complex brain surgery.

Musk is promising Neuralink can do more than science has proven: everything from addressing brain injuries to treating depression, obesity and perhaps even schizophrenia – and even one day helping us all meld with AI to protect our futures from the singularity. If he can't, rivals are stepping up to ensure someone else will. BCI became a startup market, raising hundreds of millions of dollars in funding, only *after* Musk showed the idea attention. That could well be his only legacy when it comes to BCI.

BCI was so hyped that even Facebook got in on it. Founder Mark Zuckerberg announced similar-sounding research in 2017, again aiming to meld us with machines because we won't be able to keep up with AI. He compared the data output of a human brain to the equivalent of four HD movies every second. 'The problem is that the best way we have to get information out into the world – speech – can only transmit about the same amount of data as a 1980s modem,' he said at a Facebook conference that year. 'We're working on a system that will let you type straight from your brain about 5x faster than you can type on your phone today.'

Let's walk through that. Think about how long it takes to write a sentence that actually matters, figuring out the best way to get your idea across and considering tone and audience. Now imagine what would happen if the first thought in your head was immediately broadcast. This is, of course, how some people treat social media, but perhaps Facebook would benefit from more thinking before posting, rather than the other way around. But Facebook has reasons to want into our heads. 'Eventually, we want to turn it into a wearable technology that can be manufactured at scale,' Zuckerberg explained. 'Even a simple yes/no "brain click" would help make things like augmented reality feel much more natural.' Ah, so it's all about Zuckerberg's love of the metaverse after all – though I'm not going to have brain surgery just to make his virtual world work.

The head of this project, or at least the one tasked with announcing it, was Regina Dugan, who has a remarkable CV. Formerly the head of DARPA, she set up and oversaw the Advanced Technology and Projects (ATAP) development lab at Google, before jumping ship to Facebook. Her first public words on stage as a Facebook employee were these: 'What if you could type directly from your brain?'

'It sounds impossible,' she continued, taking another measured step across the stage. 'But it's closer than you may realise.' She promised at the conference that a system to let people type with their brains three times as quickly as a smartphone keyboard would be available within three years, showing a video of a woman with ALS as an example of why this technology is beneficial. Alas, Facebook has since given up its neurological research.

We can already type directly from our brains – just very slowly. The fastest to date is a patient with ALS who topped a rate of 62 words a minute, well above what most people using an implant to communicate can manage. Indeed, that's three times the rate of the previous record. Of course, anyone who spends much time at a keyboard can easily top that: we have excellent brain–computer interfaces via our hands, after all.

Indeed, g.tec, that Austrian med tech company that let my colleague spell out his greeting with just a bit of concentration in a busy show hall, started working on its BCI headsets back in 1999. 'At the beginning, you were just controlling cursors on a computer screen, so we could just do left and right,' co-founder Dr Christoph Guger tells me. That was followed by the intendiX – the model my colleague and I saw at CeBIT back in 2010. 'This was the first brain–computer interface you could use to spell.'

IntendiX used eight sensors on the exterior of the head (no implants here) and after four minutes of calibration – a breakthrough given rival systems took a week to calibrate, says Guger – anyone could spell relatively quickly by choosing individual characters and numbers, though it also had a predictive text system to make it easier to form sentences. A g.tec researcher later improved the system by adding black and white faces of famous people to each of the letters, as thinking of a person produces a bigger brain signal response than a letter symbol. 'It was much easier for intendiX to measure it, and it gave users perfect accuracy,' he says. 'But nevertheless you needed a couple of seconds to select one.'

At the time, intendiX cost about €10,000, making it unaffordable for many who would have dreamed of such a system. So g.tec worked on a cheaper model that now costs €990 and is called Unicorn Hybrid Black – yes, really – that is smaller, with integrated circuits that send the EEG data wirelessly to a computer to run analysis, rather than doing it all on the headset. That price point makes it much easier for patients to just buy without seeking health agency or insurer approval, and it's also cheap enough that makers and gamers and artists have bought them, too, holding BCI hackathons. 'People … work together to come up with new innovative BCI applications,' he says. 'For the medical field, or for gaming, or just for data science.' The most recent hackathon pulled in 16,000 attendees, with teams quickly building projects, including a game where player one used a traditional keyboard to control the avatar while player two used a headset to destroy obstacles and clear the way for the avatar.

While CeBIT is long gone, g.tec still brings its brain-reading technology out to other tech shows. 'For many people, the technology is still new,' Guger says. 'But it's already 25 years old.'

We've had 25 years to make these headsets available to everyone with locked-in syndrome and to study their applicability to rehabilitation for strokes and MS. We've had BCI implants for decades, slowly marching through experimental studies. Yet both variants of BCI remain niche enough that people still think they're new – and novel enough to form startups around implanted versions. It's a shame we had to wait until Musk showed attention before these ideas could get the funding they clearly needed. The first BCI brain implant was in 2004 and we've had only a few dozen more since then. Part of that is down to the complexity of the surgeries and systems, but if Musk's fear of AI means more people get the help that Matt Nagle received decades ago, that's nothing but a good thing. Critics are right that Musk is simply replicating their work from decades ago, but perhaps it's also true that sometimes scientists need to speed things up a little, and if a tech bro billionaire is what it takes, so be it.

For the time being, forget embedding AI into your brain: the future of BCI should be and will be a tool to treat the worst diseases and give a voice or movement to those without, just like their predecessors' cochlear implants, pacemakers and bionic prosthetics. Regulators will hopefully ensure that these technologies keep their rightful place as treatments rather than for playing games or exploring the metaverse, but on the flip side, it's helpful for everyone developing these technologies that companies like Neuralink spark a flood of funding. If private companies realise there's no money to be made and pull back, as Facebook has, hopefully governments will step up and invest in these life-changing and life-saving ideas.

But even now, the people who most need smart prosthetics still don't get them. We have solutions to improve people's lives, we just need to be willing to pay for them. We don't need these technologies to become cyborgs, just to feel ourselves again.

CHAPTER SIX

Flying Cars

Where's my flying car? That plaintive whine symbolises the broken promises made about the future, alongside jet packs and robot butlers. But in fact, the first flying car was certified by the predecessor to the US Federal Aviation Authority (FAA) in 1950, a handful have been listed for sale on eBay, and hundreds have started the certification process in recent years. We've had flying cars for several decades. They're just not, I'm sorry to say, for you (unless you have a pilot's licence and an over-full bank account).

One example of the recent flock of flying cars is set to take to the air above Spanish olive groves, at a test flight centre in Andalusia, three hours north of Málaga's beaches. The experimental jet is parked in the already hot morning sun on the baking tarmac, awaiting commands from a team of six working from a monitor-filled trailer alongside the runway.

There's no pilot on board – there's no room, as the fuselage is packed with batteries. Instead, a former F18 fighter pilot is at the controls remotely, in the trailer, eyes fixed on dual computer monitors with joysticks in his hands. The engines roar into life, though the sudden sound is less like a lion and more the whirr of a loud hand dryer in a public lavatory. The jet, this one made by German startup Lilium, wobbles off the ground before floating confidently skywards, hovering for a moment and then swooshing off quietly and smoothly over the tops of the olive trees.

This mini jet is perhaps not what you imagine when you picture a flying car. Maybe it's the commuter saloon in *The Jetsons* that comes to mind, with its bubble-domed retro-styling, or the drive-and-fly Spinner police cruisers in *Blade Runner*. In reality, flying cars are far from such cinematic creations, though there's plenty of video: search YouTube for 'flying car' demos, and no matter how slickly produced the cinematography, your expectations are likely to come crashing down to earth. Test flights are short, with aircraft hovering briefly before plonking indelicately back down.

So what even is a flying car? For some, it's a vehicle that can be driven on roads to the local airfield, where – after a few quick modifications, such as bolting on wings or folding them out – it can then take off. In other words, a roadable aircraft. But there's another vision for the flying car: personal flying machines meant to leave roads behind entirely, hovering craft able to dip and dart through urban landscapes.

Electric propulsion has sparked a resurgence in small flying machines, but this technology has long been available. So why don't we all have air-cars or roadable planes – or the other names these have been called through the years – parked in our driveways? It's because, for most of us, flying cars don't make sense. Think all other drivers are idiots? They are, but now add height and speed to the equation, for those with a pilot's licence at least. Frustrated by road traffic on your commute to work? Flying cars don't end congestion, they extend it from flat roads to multidimensional skies, filling the air over our heads with spluttering small planes. Don't like cars clogging streets? Where are you going to park even bigger, winged vehicles? Plus imagine the noise complaints. And consider the compromises necessary to make a vehicle that can fly and drive: on the road they're slow and cramped; in the air, they're inefficiently weighed down by heavy car components.

So forget the flying car for personal use. It's a disappointment, I know, but don't despair: these new electric machines still have the chance to soar above such concerns with smarter business models, perhaps introducing a slice of sustainability into aviation.

The new generation of electric aircraft – many of which we might describe as helicopter-cum-planes – is more likely to usurp helicopters and disrupt short-haul air travel, rather than give everyone airborne commutes from their own front door. This idea has been extrapolated into flying taxis to avoid city traffic, though others see them more as flying minibuses, linking up places unserved by mass transit, or as a private vehicle for ferrying wealthy people about, the way helicopters do now. That's the near future. One day these designs may well trickle down to smaller aircraft at a price point affordable to a wider range of people, but for the moment, set aside fantasies of a soaring two-seater that's as easy to drive as an automatic car or perhaps even flown autonomously. Instead, picture a small plane with wings stretched out either side being heaved into the air by dozens of tiny rotors, a Frankenstein mishmash of drone, turbo-prop plane and helicopter.

If these are simply redesigned helicopters or planes, why are they even called flying cars? Blame journalists desperate for compelling headlines that you'll click (my sincerest apologies, this includes me too), desperately trying to avoid the dull confusing acronym favoured in the aerospace industry. Those in the field call such vehicles 'VTOLs', which stands for Vertical Take-Off and Landing, as that's exactly what these vehicles do. Like a helicopter, they don't need a runway, just a helipad for going up and coming back down. Unlike a helicopter, some also have wings for more efficient flight, hence the designs that look like the offspring between a chopper and a turbo-prop plane.

But eVTOLs – the electric version, as batteries have spurred the recent boom in development – are just one design in a long line of so-called flying cars wobbling into the skies over the last few centuries. We've seen planes and cars bolted together, roadable aircraft with folding wings and even homemade wobbling spaceships. Flying cars aren't the future, they're decidedly retro.

★ ★ ★

In 1772, Abbé Desforges, the canon of Étampes to the south of Paris, France, clambered to the top of the town's tower with an associate and a wheeled, winged flying machine dubbed a 'cabriolet'. They pushed off, hoping to soar in the air, but physics disagreed. They crashed to the ground, destroying the cabriolet and spraining Desforges' ankle.

It's easy to scoff at such efforts, so similar to a toddler strapping on self-made cardboard wings and leaping from a garden wall. But humans have been trying to fly forever, at least all the way back to Leonardo da Vinci's sketches in the 1480s, though it's safe to assume the urge began much earlier. When the Wright brothers finally evaded gravity and wobbled into the air at Kitty Hawk, North Carolina in 1903, it was amid a global race to take to the skies in flying machines, with rivals including Samuel Pierpont Langley, Glenn Curtiss, Alberto Santos-Dumont, Louis Blériot, and many others. At the same time, cars were starting to arrive on city streets. In 1885, Carl Benz made what's widely considered the first automobile, with Henry Ford founding his Detroit Automobile Company in 1899. Personal mechanised transport had arrived, as had human flight. It made sense to smush together these two technologies, using newly developed engines to power personal flying machines that would have let Desforges soar rather than stumble.

There was plenty of stumbling by engineers as creative and talented and bonkers as the Wrights, but just a bit less lucky. A year ahead of the Wright brothers' successful test, Romanian inventor Trajan Vuia travelled to France to find support for his monoplane designs, submitting his 'aeroplane-automobile' to the Académie des Sciences in Paris. They rejected it as an impossibility, saying 'the problem of flight with a machine which weighs more than air cannot be solved and it is only a dream'. Thankfully he ignored those wrong-headed academics and set to work building. His four-wheeled, lightweight craft looked like a steampunk cross between a bike and a moth, with a propeller at the front, and rear rudder for horizontal balance. The contraption used a single set of wings, rather than the more common biplane design with stacked wings – after all, Vuia noted, 'I have never seen a bird with more than two wings.'

At first Vuia only drove it, wingless, as a car; but in 1906 he slapped on the wings and flew 12m (40ft), at just a metre high, before engine problems grounded him. That may not sound like much, but in doing so it became the first aircraft to take off from a flat surface, without the benefit of a ramp or external help such as the Wright brothers' launching rail. Over subsequent flights, Vuia altered the design, eventually filing patents for automobile aeroplanes with foldable wings, but a crash eventually damaged the craft and lightly injured Vuia. Given his lack of success sustaining flight, he stopped testing it; 15 years later, he developed early versions of helicopters.

Vuia wasn't alone in seeing the links between the car and the plane. In 1909 Russian Vladimir Tatarinov unveiled his Aeromobile, a car with a propeller at the front and four six-blade rotors – a design that doesn't look too different from flying cars today. He was supported using military funding, also similar to many companies today, though they did impatiently cut his funding, so we don't know if it would have worked.

In 1917, a rival of the Wright brothers, Glenn Curtiss, unveiled his Autoplane at the New York Pan American Aeronautic Exposition. Curtiss was taken seriously because he had serious credentials: a man of many firsts, he was the first American with a pilot's licence (and the second in France). Nicknamed the 'aerial limousine' by the media, his aluminium aircraft looked like an aerodynamic version of touring cars of the era – which is to say it looked a bit like a roller skate for giants. It was driven on the road and in the air by a pusher propeller, located at the back like a motorboat propeller to pull air over the wings. The 12m (40ft) wings and tail were removable. Though it was roadworthy at 70km/h (45mph), the

Autoplane only ever managed short hops, and a lack of investors led Curtiss to refocus his efforts on planes that could land and take off from water, where he found success, with 100 of his flying boats manufactured by the US Navy.

Perhaps the most successful early effort came in 1921, with René Tampier's Avion-automobile. Rather than the wings needing to be removed after landing, in Tampier's clever design they could be folded back to hit the streets of Paris, which he did in November of that year, to star in the Salon de l'aviation (Air Show – see, everything is better in French). Built as a war machine rather than a commuter craft, the Avion-automobile's arrival after the First World War, rather than during, rather limited its potential. Timing is everything.

Another design for roadable aircraft came via the autogyro. Created by Spanish inventor Juan de La Cierva, it's a bit of a mix between a glider and a helicopter: an unpowered rotor on the top spins because of air passing over it, giving the craft lift, while powered propellers offer thrust. The first successful flight was in 1923, and it was a design that not only sparked the invention of helicopters but became popular among roadable aircraft builders, inspiring copycat designs from American Harold Pitcairn, who licensed the Spaniard's designs. In the 1940s, the British army tested a version called the Hafner Rotabuggy as a way to drop all-terrain vehicles off at tough-to-reach locations; the design was never used, however.

In 1926, Ford unveiled the Flivver.* Despite the recognisable Ford logo painted down the side, and headlines proclaiming it 'Ford's Flying Car', it was more of a cheap plane than a flying car, intended to bring affordable flight to the masses and required to be small enough to fit inside Henry Ford's office. The small team led by Otto Koppen developed three or four Ford prototypes, all flown by the same pilot, Harry Brooks, the son of Ford's friend. Brooks was by all accounts a

* The word 'flivver' was initially used to refer to cheap cars, but was quickly applied to planes, too.

remarkable pilot – his skill is perhaps why the aircraft looks so smooth in take-off in online videos – and highlighted the use of the Flivver as a commuter craft by using it to get from his suburban home to work at Ford as well as to golf, landing directly on the course.

But in 1928, the then 25-year-old Brooks took the Flivver to the air over the coast of Florida where it crashed into the waves, sparking an immediate rescue operation. Witnesses reported seeing the pilot standing on the wreckage, and the safety straps were later found unbuckled when the plane was eventually found. Brooks' body was never recovered, though his wallet and a thermos bottle washed up ashore. Aviators flew over the spot where the plane crashed to drop a memorial wreath from 300m (1,000ft) up in tribute. Devastated, Ford lost faith in the project and shut it down – but this didn't stop his belief in flight for everyone. In 1940 he was quoted as predicting: 'Remember what I'm going to tell you: a combination of plane and car will see the light of day. You may smile, but it will come.' He just wasn't going to be the one to build it.

⋆ ⋆ ⋆

Waldo Waterman didn't intend to build a flying car. Indeed, no one was really sure what his first creation even was when they looked at it: a tailless aeroplane with a rear propeller, it was designed to be easy to drive and landable on a road. The press kept asking 'What is it?', so he named his craft the 'Whatsit'. Five years later, he followed up the Whatsit with the Arrowplane, built to enter into the 1935 Vidal competition. This was a challenge run by the chief of the US Bureau of Air Commerce, Eugene Vidal, who asked aviation engineers to build a 'safety plane' or 'baby plane'. It was specified that it should cost $700, about the same cost as a car, and be as easy to drive, in the hopes of creating a 'nation of pilots' who could fly just as easily as drive.

After none of the designs came in near the desired price point, the government abandoned its plans to mass-produce cheap planes, so Waterman adjusted his design to become the roadable Arrowbile. It had detachable wings, a Ford radiator and a Studebaker car engine for both flight and driving on the road, where it managed 180km/h (110mph) and 90km/h (55mph) respectively. The three-wheeled configuration let it be licensed as a motorbike. Studebaker ordered five production models, and three were made;* the first crashed, the second had mechanical issues and the third broke down, nearly killing the pilot. That was a problem as it was required for a press event nearby, but a local farmer helped drag it over and put it back together before the media arrived. Despite such efforts, there was little interest from the market, thanks in part to the high cost of $7,000.

The arrival of the Second World War pushed back efforts to build commercial planes. But the subsequent return home of a swathe of newly trained pilots, as well as the return to light aircraft production after being banned in the US during the war, and more money in everyone's pockets was expected to spark a boom in personal aviation. At the same time, the postwar optimism paired with the push for modernity spurred the idea of 'tomorrow's world', where commutes were airborne and weekends away weren't limited by slow motorways. By 1947, there were more than 400,000 licensed pilots, a third of them on private licences, with 17 companies selling 30 different aircraft models. You could even buy them in department stores. In 1945, Macy's began stocking the ERCO Ercoupe, hiring two former military pilots as salesmen and taking out a full-page ad in the *New York Times* saying the plane was 'as easy to handle as your family car'. Other retailers quickly followed suit, also stocking the Piper Cub. While

* A fourth was made many years later, making a test flight in 1957. That model is now in the US National Air and Space Museum.

orders initially piled up, it soon became clear there was a ceiling to personal aircraft sales.

Macy's stopped selling planes two years later, offering the display model for a discounted price of $2,100, with only 5,500 Ercoupes having ever been built. Production of light aircraft topped 33,000 in the US by the end of 1945, but by 1947 sales had slid to 15,000, and two years later to 3,500. The economy hadn't taken off as hoped, jobs weren't easy to come by and there were plenty of other costs to cover. Planes, roadable or not, were a luxury few could afford.

★ ★ ★

As they stood and watched years of engineering work, money and a pilot crash into the mudflats near San Diego Bay, Theodore Hall and his colleagues learned the hard way that roadable aircraft were a tough sell even when they worked.

Hall began working on a roadable plane in 1939. During the Second World War, he had the notion to use flying cars for air raids – like many new technologies, the flying car sprung from an urge for military might. His employers at Consolidated Vultee Aircraft (Convair) even believed they could mount a machine gun on the front.

But after the war, Hall quit Convair to focus on the idea, eventually developing a prototype dubbed the Hall Flying Car. It was very much a car, featuring an engine used by the popular carmaker Crosley.

There was plenty of interest. Hall wasn't seen as a bonkers inventor. Aeronautical companies asked for his help designing their own models, including Texas-based Southern Aircraft Corporation; its flying car tested well, but never quite made it into production. Indeed, the Hall Flying Car was so intriguing to industry that Hall's old employers at Convair came knocking. The company was already working on small personal planes, and had on staff William Stout, who had worked with Waldo Waterman.

The first version, the 116, was a two-seater that ran 66 test flights; we have plenty of detail about it because Hall invited journalists from *Popular Mechanics* to take a look, though the long-running magazine gave the flying car just a single page of coverage, as much space as a story about the welcome return to metal toys post–war.* According to that report, the 'roadable plane' designed by Hall topped 180km/h (110mph) in the air and 100km/h (60mph) on the road, though it couldn't just be landed and driven on to the highway. The propeller, 9m (30ft)-long wings, and tail first had to be removed, leaving behind a three-wheeled car in a standard chassis. The engine crankshaft was used to turn the propeller using a standard car transmission, and there was space in the boot for luggage.

The second model, the 118, added an upgraded chassis and more powerful engine. It really was a car and a plane smushed together – it looked like a classic sedan with a plane glued on top, but in a good way. The car aspect was designed by Henry Dreyfuss, a famous industrial designer who was formerly a student of Norman Bel Geddes, whom we met earlier with his ideas for driverless cars. Dreyfuss knew how to make iconic objects: Bell telephones, John Deere tractors and Polaroid cameras look the way they do because of him.

The Convair (or ConvAirCar) 118 was a car with an attachable plane. The business model they imagined was to sell the cars but rent the plane component, which customers could pick up at the airport. To assemble this transformer of a machine, the plane was propped up by a tripod support system, and the car reversed into position underneath it. Then, the plane component was lowered down onto the car. The driver sat in the same seat, but now had the cockpit and

* According to that April 1946 issue of *Popular Mechanics*, 'bomb scooters' and 'pedal bombs' were the hot toys that year, which were metal bicycles in the shape of bombs. For the girls, or 'future wives' as they were called in the article, there was a toy sewing machine.

other controls in front of them. Convair was bullish: the company had a production target of 160,000 cars sold at $1,500 each, plus the rental plane bits.

One test flight, in November 1947, saw the ConvAirCar 118 circling San Diego for an hour and 18 minutes – much longer than that four-minute long Lilium test flight in an empty bit of Spain 76 years later. But the test ended with a crash in the mudflats when the pilot, Reuban Snodgrass, accidentally looked at the car's dashboard rather than the one for the plane, and subsequently ran out of fuel. Snodgrass walked away with minor bruises, though flight researcher Lawrence Phillips fractured his shoulder. The lower, car part of the craft was demolished, though the top half made out of a plane was only slightly damaged. It was restored, paired with a new car, and flew again – with a different pilot this time – but the crash appears to have cast a pall over the project. It was cancelled by the company after it became clear that sales of personal aircraft were nowhere near what anyone had expected, and were already tailing off.

Others had better luck. The Airphibian was certified for flight by the precursor to the FAA in 1950, making it the first approved flying car. Developed by Robert Edison Fulton, it was a metal car with large wheels and removable fabric wings. Convertible in under five minutes, the Airphibian was intended as a mainstream vehicle, and reached 80km/h (50mph) on the road and 180km/h (110mph) in the air.

Beyond building the world's first certified flying car, Fulton held more than 70 patents, with inventions ranging from the first aerial flight simulator to the skyhook, a tool used by the US military to pick up stranded people using aircraft. When he passed away at 95 in 2004, the *New York Times* described Fulton as an 'intrepid inventor', an excellent and accurate epitaph. Though successful in so many other aspects of life, the costs of building the Airphibian meant Fulton had to sell his designs to a company that never brought it to production. Perhaps that's no surprise given a core flaw replicated by many of these craft: it could fly and it could drive, but it wasn't particularly good at either, with one review calling it 'underpowered as an airplane and overpowered as a car'.*

But another roadable aircraft builder: Moulton Taylor, was inspired by Fulton's work to create his own vision, the Aerocar. Taylor earned his pilot's licence as young as locally possible, at just 16, before studying aeronautical engineering and joining the navy when the Second World War arrived, working on a project to build remotely guided surface-to-surface missiles. After the war, he decided he'd rather not continue working at military labs. 'I just decided not to spend the rest of my life making things to kill people,' he said in an interview in 1958. 'My dream is to look up and see the sky black with Aerocars – and I'm sure that will happen someday.'

Looking a bit like a toy brought to life, the Aerocar featured folding 10m (34ft)-long wings and a removable tail section, both of which were pulled by a trailer and could be installed into place quickly. A marketing brochure promised the transformation would take just three minutes, though one assumes that was with a bit of practice. To complete the conversion, the driver-pilot flipped up the rear licence plate to pull out a shaft to attach a propeller, driven by the same engine as in car mode. 'The changeover involves about as much work as changing a tyre,' the presenter promises in an episode of *Industry on Parade*.† 'As an automobile, its appearance is rather unusual, but as an airplane, it looks pretty much like conventional models.' On the road, the Aerocar could drive at 100km/h (60mph) – plenty fast enough for the time period – and in the air could fly at 180km/h (110mph).

---

* The last remaining Airphibian is on display at the Smithsonian: https://s.si.edu/3RxqtG5

† You can watch the Aerocar in action here: https://biturl.top/JVjyeu. 'We called it a dream of the future, but actually the Aerocar is already in production,' croons the presenter. You can also see one on display at the Museum of Flight in Seattle.

Taylor won certification for road and air travel in 1956 and signed a deal to produce the Aerocar, but only if he could sign 500 orders. Sadly, he only managed 278, perhaps because of the high price of $15,000, at a time when a new plane or new family car could be bought for $2,000 each. In the end, just six Aerocars were built, but some of them did find fame. Actor Bob Cummings used one Aerocar for his 1960s eponymous TV show, while another was used by a radio station for traffic reporting – much cooler than a helicopter. One of the Aerocars was flown to Cuba, where the owner gave a ride to Fidel Castro's brother, Raul, though it was damaged by a spooked horse after an emergency landing on a rural road. There are no reports as to whether the horse was okay or not, but let's choose to assume it won that battle.

Taylor didn't give up on the dream, at one point nearly signing a deal that would see Ford manufacture Aerocars, but this was stymied by the Department of Transport requiring that the vehicle be certified not as an aeroplane but as a car, a change that would have cost too much to manage. That was followed by plans for a flying car conversion kit that would pair a Honda CRX with towable, removable wings and tailpiece that would hold its own engine and propeller. The aim was to make a car that people could assemble at home, though it was never built, as Taylor passed away in 1995. It seems like Taylor never stopped believing, though, telling an interviewer: 'If I can get this thing flying, the world will beat a path to our door.' He's not the first nor the last to think that.

The 1940s saw an explosion of flying car or roadable plane ideas. There was the York Commuter, a small car that you'd drive up the ramp into the cabin of the aircraft, where it connected into the controls so the driver could fly without changing seats. Similarly, the Aviauto was a plane with a modular fuselage that could be swapped to different designs, turning the vehicle into an ambulance, freight transporter or a five-seater flying car; while that meant lugging a car engine into the air, it was at least used to power the inflight radio.

Luigi Pellarini built the Aeronova, with wings that not only folded back but pivoted to stow vertically, while the plane propeller was used to help drive the car. Leland Bryan also built multiple models that used the propeller not just to fly but to drive, while Henry Smolinski melded a Cessna Skymaster to a Ford Pinto – I'm afraid this experiment ended sadly and only reinforced the American-made car's reputation as a fire risk, as it burst into flames, killing its creator. While tragedy did strike, there was no shortage of designs, and plenty of them actually worked – though nearly all of them had ridiculous portmanteau names smushing 'flying' and 'car' together. Being good at engineering doesn't make you a great marketer, it would seem.

Flying cars were approved by regulators and ready for production 70 years ago. And there were plenty of other designs between then and now, with 68 patents filed in the US alone between 1906 and 1994. So why are startups racing to build them once again? Why did previous efforts fail? *Where's my flying car?*

Costs are one reason: for the price of an Aerocar, you could have bought a new car and a new plane and still had $10,000 left in your bank account. Another reason is the danger. Henry Ford ditched his foray into aircraft after the death of his test pilot, Hall's Convair stalled after a single crash, Leland Bryan was killed in the 1960s by a wing installation failure, while Henry Smolinski and a pilot died in 1973 testing that Cessna/Pinto.

Perhaps, with some designs more than others, roadable aircraft or flying cars weren't very good at either task – who wants to drive a terrible car just to be able to fly a terrible plane? Or maybe our constant query of *where's my flying car?* doesn't actually translate to real demand. We might keep asking for flying cars, but that doesn't actually mean we want one. That still hasn't stopped people from trying, though.

* * *

Paul Moller is the epitome of single-minded: for seven decades, he has methodically designed, built and tested his flying saucers and roadable aircraft after a first flight in 1967 – despite a fire wiping out his workshop, an investigation by financial regulators and falling so low on cash he tried to sell one of his prototypes on eBay.

Indeed, Moller has been thinking about flying cars for even longer, his interest sparked when he rescued a hummingbird as a child, imagining it as 'a great way to get to school'. Even as a child, Moller was an engineering prodigy, designing and building his own small house, a working Ferris wheel and, by 15, even attempting a helicopter, using the kitchen stove to bend the plexiglass for the windscreen. 'I certainly didn't get very far,' he tells me, though the aircraft got some use from his father, a chicken farmer, who used the propeller to keep incubated eggs cool when temperatures got too warm.

Moller's first flying car didn't lift off into the air until he took a role as a professor at the University of California, Davis, but he worked on it throughout his student years. 'When I did my PhD at McGill University, I had this project going on the side – like I didn't have enough to do with my PhD,' he tells me. Then he moved from Canada to California for a professorship at UC Davis. 'I would teach my students ... and then I would come home and work until dark.'

He built the first two aircraft in his own time, but eventually the university sponsored him to make a few improvements and run a test flight in 1967. The M200X was a 4m (14ft)-long flying saucer with no stabilisation, so it bounced and rolled like a bronco – 'rocky, but great fun', he said in a documentary of his work.* It had to be held down while Moller started the engines, and because it hovered just 45cm (18in) off the

* The documentary, *Father of the Flying Car,* is available online. There's also a short video of the flying saucer here: https://youtu.be/4cvQ2nLXFw

ground it raised such a dust cloud that he once ran into a parked car.

The dusty crash didn't put Moller off – that, it would seem, is impossible – and he continued working, refining the design and creating a stabilisation system. In 1989, he flight-tested a two-seater, saucer-shaped VTOL with a circle of engines along the bottom. Videos of two test flights show it slowly rising off the ground, the grass beneath him ruffled by the engines, the aircraft wobbling in the air in an unreal manner.

Moller flew the VTOL himself, well aware that technical difficulties – in particular, a known risk with the blades separating from the fans – could kill him. 'I don't think I could be more elated than when I realised after two minutes in the air I was going to live,' he said. Beyond relief, he also felt wonder at the 'magic carpet' sensation of being lifted up from below rather than pulled up from above. If successful, the M200 Neuera promised a cruising speed of 120km/h (75mph) and a maximum speed of 160 km/h (100mph), but only 160km (100 miles) of range. It was 3m (10ft) across and 1m (3ft) wide, and designed to carry just one person.

That aircraft wouldn't necessarily work for a commute, though if taken to market, Moller believed the M200 Neuera could work as a recreational vehicle, or be put to work in border control or fire and rescue. One investor-customer wanted to use it to patrol their farm. Instead, Moller changed tack, reconfiguring the design into more of a flying car, or roadable plane, called the Skycar. With the Skycar M400, its four engines point at the ground for take-off before rotating into position to fly – a design that's seen on eVTOLs today. Bright red and sporty, it certainly fits the sci-fi aesthetic, looking like a space plane for superheroes. The aim is for it to one day be autonomous, but that is some time off. In the meantime, prototypes are designed to be simple to fly.

The work was expensive. To fund it, Moller sold off his successful motorcycle muffler company and his research park, and twice tried to sell prototypes on eBay. There were no takers, and crowdfunding efforts also failed. He began taking $5,000 down payments from his ardent fans and sold stock on the company, triggering an investigation by financial authorities. Then he declared bankruptcy, and his workshop burned down. He switched focus again, shelving the Skycar in favour of a return to the Neuera flying saucer, which Moller believes will be easier to get certified and into the air. The work has been facilitated by a younger generation of workers, who have upgraded the analogue-era controls with digital components that are much easier to use, lighter and cheaper.

Such technological leaps forward have made life easier for anyone developing VTOL-style personal flying machines now, as it's like not having to invent the wheel to make cars. But Moller isn't convinced by the current crop of eVTOLs – not even those developed by Joby Aviation, a company founded by his former employee JoeBen Bevirt – in particular, because he has doubts about battery-powered flight.

He's now hoping to release the Neuera as a recreational craft, dodging some of the toughest certification challenges. That's not because he's concerned the flying saucer wouldn't pass, but simply because regulatory timeframes are long – and he's not getting any younger. By making a single-person craft, he hopes to speed up the process. 'The FAA doesn't get so nervous about you killing yourself, right?' he tells me. 'You have a one-person vehicle, you're taking your own life, [in] your hands, no problem.'

Others have also believed in the flying car dream, though have lacked the funding to get as far as Moller. In 1994, Ken Wernicke's AirCar combined a VTOL with a roadworthy car, though it looked more like a retro space plane, and it never progressed beyond the model stage. In 1997, Steven Crow unveiled the Starcar 4, with wings that folded up on the sides. It never went into production. And in 2005, Carl Dietrich revealed the Terrafugia Transition, an aircraft with foldable wings that could be driven on roads. Unlike the others, the

Terrafugia was actually flown in 2012, after receiving a special licence for light aircraft, despite being technically too heavy because of the necessary safety features required by a car. The base price has risen from $194,000 to more than $300,000, though no production models have been made, despite deliveries having been promised for 2012. After the company was bought out by Chinese investors, Dietrich departed and all US operations ended. It's hard to get flying cars off the ground.

★ ★ ★

It's 35°C (95°F) in the middle of the Andalusian hills, though it's only 8.30am. Rows of olive trees stretch out to the horizon and beyond in every direction from where we stand on the roof terrace of the ATLAS flight test centre, the Spanish outpost where Lilium tests its experimental electric jets. One is parked on the tarmac below, bright white, shining in the heat. Christophe Hommet, chief engineer, warns the small group of journalists to speak quietly, so as not to disturb the control room directly below us in a canvas-covered trailer. 'Don't laugh at Daniel's jokes,' he says, laughing at his own made at the expense of co-founder and former CEO Daniel Wiegand.

It's so quiet we can hear the tweeting of small birds darting through the hot air. And then Hommet gets a call: the control tower behind us has spotted a fuel truck on a road amid the olives somewhere we can't see, delaying the test flight. Unlike driverless cars, tested around real people (and their pets), these flights won't run if anyone is nearby – making these Spanish hills an ideal location for an aviation test centre, as the weather rarely interferes and there's nothing but trees for miles.

The truck eventually moves along, meaning the Lilium jet can take to the air. This isn't the production model eyeing certification, but an experimental demonstrator vehicle used for testing new ideas – and demo flights for journalists. These demonstrators are officially known as the Phoenix model, but each individual plane is given a name, female of course. May

I introduce you to Lucy? This is her 101st flight, and she's smaller than the final Lilium Jet design, with enough space for maybe two or three people, though there are no seats, just controls and batteries filling the interior. The final version will see smaller, more power-dense batteries designed into the exterior body of the jet. But otherwise, Lucy is the spitting image of her larger sibling, with her sleek and curved body with four wings jutting out: two short ones at the front, and longer limbs extending at the back.

The wings are in two parts, with the back half jointed and moveable, the wavy white chassis concealing dozens of engines. This is the magic of eVTOL: those engines point at the ground for take-off, pushing the jet into the air. To move forward, the engines transition in line with the wing, their angle decided by how fast the aircraft needs to move. Beyond vertical take-off and landing, this allows aircraft like the Lilium Jet to fly incredibly slowly, which is helpful for navigating cities. My iPhone videos of Lucy coming in to land look as though she's flying in slow motion, but she's half-hovering, half-flying.

Of course, we're not in a city...the only thing that's high density here is the groves of olive trees. And though eVTOLs like Lucy are slowly marching their way towards regulatory approval, ensuring urban dwellers and city authorities are happy with the electric whirring over their heads is another thing altogether. Most major cities don't allow much in the way of private aircraft. Companies such as Lilium are not just building new forms of air travel, but disrupting the skies above us – the former might be easier to achieve than the latter.

The aircraft themselves are nothing short of remarkable, pairing well-established flight principles with new technologies such as electric drivetrains and autonomy. A VTOL's journey has two modes: take-off/landing and flight. Take-off is similar to a helicopter, where the spin of the rotors creates lift using pressure caused by different air speeds on the top and bottom of the wings. Think of the spinning rotors as tiny wings that lift the vehicle from the ground.

Now you're hovering in the air, it's time to fly. At its most basic, to fly forward, a helicopter tilts in the direction the pilot wants to go. The same forces that pulled it upwards now pull it forward. However, that's not a particularly efficient way of flying. Once airborne, planes are more efficient – on short flights, a quarter of their fuel can be used in take-off alone. A plane's engines simply pull the plane forward, while the wings do the hard work of defying gravity and keeping the craft in the sky. Wings alone can be enough to keep an aircraft going – gliders don't have engines, after all.

Combining the take-off of a helicopter with the efficient in-air lift of fixed wings is, in theory at least, the best of both worlds. No wonder the idea has appeal – and it does. The Centre for Future Transport and Cities at Coventry University tracks eVTOL aircraft concepts in a database. As of 2022, more than 500 concepts had been unveiled, though fewer than a third had yet to actually fly.

Most of the designs Coventry tracks for urban air mobility (UAM – sorry, the industry loves these acronyms) are VTOLs that are electric- or hybrid-powered, seat fewer than 10 passengers and have a take-off mass of less than 3,175kg (700lb). But there are some differences, in particular between wingless and winged designs.

In the wingless category, there are multicopters – these can look a bit like mutant helicopters or massive drones. They are effectively electric helicopters, but their designs have been tweaked with the aim of making them lighter and cheaper and more manoeuvrable than a traditional helicopter. One example is the German-built Volocopter. The VoloCity design has a two-seater helicopter-style body,* but instead of the familiar rotor swinging in a circle overhead, it has a fixed

* The first two models, the Volocopter VC1 and VC2, didn't even have a body to house passengers or a pilot. Instead, the VC1 had a seat for a human surrounded by arms on which sat tiny rotors. The first test pilot clearly had nerve.

circle supported by six arms, on which 18 propellers hum away. It sort of looks like a helicopter's rotor arms have jammed in place, on to which tiny versions have grown like propagating plants. Others include the American-made Jaunt Air Mobility Journey and the Chinese-built EHang 184 Autonomous Aerial Vehicle (AAV). The 184 is designed to be fully autonomous, and the company's test videos in February 2018 show the CEO taking a ride and living to tell the tale.*

The other category of eVTOL aircraft have wings, like the Lilium Jet, so can cruise like a plane – but only once they're in the air. To get up there, they require one of three different types of lift system: vectored thrust, independent thrust, and combined thrust. With vectored thrust, the spinny bits that provide forward propulsion also provide the lift, tilting at cruising altitude to change roles. This is what Lilium's Jet and Joby Aviation's eVTOL do. With independent thrust, there are two separate systems: one does the lifting, one does the forward flying – this is the system used by Wisk, a company spun out of Kitty Hawk, a flying car company co-founded by Larry Page and Sebastian Thrun. And with combined thrust, some engines provide lift, and some rotate to do both; this is the model used by Vertical Aerospace, and these tend to look like planes with drones stuck to the wings.

Because of the features and trade-offs of each, they're better suited for different roles: in particular, flying within cities or between cities. This is a bit of a generalisation but without wings, a quadcopter hovers well and offers stable flights in windy conditions, but has a slower flight speed, so it's best for shorter, in-city flights – they're good replacements for helicopters and could be used for air-taxi services. Designs with tilting or fixed wings can fly faster and more efficiently, making them ideal for intercity flights or as airport-to-airport shuttles. But that's a really loose rule of thumb.

---

* The EHang flight video is here: https://biturl.top/Jn26by

Undoubtedly, the reason behind the sudden surge in VTOL development is the addition of that 'e' for 'electric', though some of the companies are looking at hydrogen or hybrid designs. The development of electric drivetrains and significant improvements in battery technology made eVTOLs possible, while the impact of burning jet fuel on climate change is extra motivation to shift to more sustainable flight.

When entirely electric, there are no operational emissions, which means there's nothing nasty coming out of the tailpipe.* A standard return flight on a jet fuel-powered aeroplane from London to Paris spews out about 244kg (540lb) of CO_2 per passenger, while the same journey via train is about 22kg (50lb).† A study in *Nature* suggested that not only could eVTOL be better for the environment than internal combustion cars, but also an improvement on electric cars, producing 52 per cent less emissions than the former and 6 per cent less than the latter. There are a few caveats, notably that for these calculations to hold true the eVTOLs need to fly at near-full capacity, while the benefits are only significant for trips above 35km (22 miles) with most car journeys seeing benefits at less than half that distance. That means flying cars could be more sustainable for longer trips, but for short jaunts around the city they're more about convenience. And, as ever, electric vehicles are only as green as the local electricity mix – if you're in a state or region that burns coal to generate electricity, this is all rather moot.

* Operational emissions are what come out of your car's exhaust pipe, but of course carbon emissions will be created when these jets are manufactured, and when electricity is produced, so Lilium is right not to make the claim that its Jets are entirely emissions free.

† That's according to figures from the delightful train website The Man in Seat 61 (https://biturl.top/vqIzM3). Results vary from different carbon calculators, which isn't helpful, but the point is that flying is crappy for the environment.

However, batteries aren't as capable as we'd like. If you play too many games, your smartphone might not last through the day without a recharge – and it isn't defying gravity with a dozen electric jet engines. In 2018, Uber laid out a plan for flying taxis – but at the very same time, admitted the required batteries don't yet exist.

Ensuring such plans are possible will take a leap forward in battery technology. Big batteries of course do exist, but they weigh a ton – well, half of one. The batteries for the long-range Tesla Model 3 weigh in at 480kg (1,060lb), but the Lilium craft's total weight is 440kg (970lb). This is the reason why electric propulsion is being used for eVTOLs and not passenger planes – with today's technology, the battery required for that would be massive and unwieldy. Lilium's Wiegand believes that will change, and hopes that once the current version of the Jet is up in the air and flying, the company can shift to developing electric engines for conventional airliners running longer-haul flights. 'They'd be limited in range, maybe 700 or 800 miles [1,100–1,300km] or so, but you'd be surprised how many flights today are less than that,' he told me. 'If we had a 1,000km (621-mile) range, we could globally replace up to 50 per cent of all flights and make them sustainable. And that's a big, big need for the planet.'

Jet fuel is toxic for the environment but it's efficient in terms of weight, offering 11,890 watt-hours of energy per kilogram, versus the 250 watt-hours per kilogram of car batteries. Passenger jets would require 800 watt-hours per kilogram for longer-haul routes, according to research by Andreas Schafer at University College London. In a paper he notes that energy densities have been increasing by 3 to 4 per cent per year, meaning we should hit the required density by 2050, if development continues at a similar rate. And that is a big if.

So we're going to need a better battery. Thankfully, plenty of alternatives are in the works to replace or supplement existing lithium-ion designs, from high-density solid-state batteries to lightweight lithium–sulphur designs. Whether any of those ideas pay off, though, remains to be seen. Alongside a better battery, improvements can be made to flight efficiency so less power is needed, and by using hybrids, such as Bell's Nexus, which pair electric with gas. Those are serious challenges, but they're not insurmountable. When the battery issue is solved, longer flights will be possible.

While batteries may well be the challenge that grounds eVTOLs in the immediate future – no one wants to take a four-minute flight, after all – the non-technical hurdles, as ever, may prove the toughest to overcome. Before eVTOLs are offering passenger flights, they need to sort out regulation and certification, safety and perception, and build the necessary infrastructure, including air traffic control.

Let's start with regulators. Often regarded as the bogeyman holding back the utopian dreams of startup founders, they're the ones responsible for ensuring air travel is safe. And regulation is not only rightly tight in air travel, but the aerospace industry is accustomed to working with the likes of the US Federal Aviation Administration (FAA) and European Union Aviation Safety Agency (EASA), who between them regulate 80 per cent of the world's aviation activity, as well as with local regulators such as the UK's Civil Aviation Authority (CAA). When those relationships go wrong, it can be catastrophic. A US government investigation found that the US FAA lacked the ability to analyse Boeing's changes to the 737 Max passenger jet, instead trusting the company to ensure the system was safe. It wasn't, and the ensuing two crashes in 2018 and 2019 killed 346 people.

All new designs, which includes the vast majority of eVTOLs, need to be certified, and that comes after a multi-year process which allows regulators to examine every aspect from production to maintenance, including a company's safety culture. So when companies say they expect to be flying passengers within two years, or by a certain date, they are giving us the best-case scenario. It's how long the process will take if nothing goes wrong and everything runs perfectly. Delays aren't necessarily a bad thing, but a sign of complexity. On the flip side, an eVTOL may be described as having regulatory approval, but that means different things: it could simply indicate that a company has won permission to test, to seek additional certification, to fly in certain areas or with people, and so on. And they don't just need approval to take to the sky: US startup ASKA also had to seek permission for its roadable aircraft to drive on public roads.

★ ★ ★

As eVTOLs are slowly flying towards certification, companies are popping up to build places for them to land, often referred to as 'vertiports'. Uber's app means you can get a car and driver at your door within minutes, but don't expect a flying taxi to be touching down in front of your house … ever. Boeing's PAV is 8.5m (28ft) wide, while Bell's Nexus is 12m (40ft) across. Unless you live on a Parisian boulevard, the road in front of your home is narrower than that, and it sure isn't neighbourly to smush neighbours' windscreens with your ride's wings. And most people don't live in high-rise buildings in urban areas with helicopter pads on the roof, though some do, and one building in Miami earned a few headlines for including a vertiport*. With increasing urbanisation it's easy to imagine such a setup becoming more commonplace.

While helicopters – and some wingless eVTOL designs – can land in any relatively flat and clear space, that's not true of all eVTOLs. As Hommet notes, the Lilium Jet can't just plonk down in a field, as the jet engines will suck up detritus from

* That apartment block having a vertiport doesn't mean the developers believe people will be moving in with flying cars anytime soon; instead, they're future-proofing it for the next 100 years. That almost sounds sensible.

the ground, causing damage. A specially prepared, and clean, tarmac surface is necessary, akin to a helipad. Other physical requirements include electric charging points and a stable grid to power them, a hangar for storage and maintenance, and plenty of room for vehicles and people.

It's easy to forget the human side of such infrastructure. For a month in Coventry, England, a company called Urban-Air Port set up a trial vertiport next to the city's railway station, letting visitors cosplay as eVTOL passengers, walking through security, waiting in the lounge and boarding a flight – all for aircraft that don't really exist yet. Renders often show vertiports atop skyscrapers, but that may prove more challenging than it looks, as rooftops house key equipment such as lift mechanisms, may require structural strengthening since as they're not designed for heavy loads, and will require access for passengers, perhaps requiring a dedicated lift. Indeed, as eVTOLs can take off much more vertically than a helicopter, they can in theory use much lower, more accessible launchpads, perhaps a multistorey parking garage or reinforced roof of a railway station.

Location matters for another reason: noise. The convenience of a local vertiport may not be outweighed by the ever-present whirring from take-offs and landings – after all, many people don't want to live by a noisy motorway or alongside rumbling train tracks. That said, eVTOLs should be quieter than a helicopter, in particular if they're fully electric and use smaller rotors. Uber wants VTOLs to be half as loud as a medium-sized truck driving past your house. Lilium's experimental jet certainly drowned out the birds at take-off, but as it flew its circuit over the olive groves it was no noisier than a drone – and that's without the sound insulation that will surround the commercial version of the craft.

A less easy problem to solve is managing a sky full of aircraft. Visual flight rules are used in some cities, and all that really means is that not crashing one helicopter into another is down to the pilots using their own eyes. That could get too complicated when hundreds of so-called flying cars congest the sky, so a smarter air-traffic control system may be needed, perhaps using AI.

And then there's perception – put another way, would you get into a new-fangled flying machine? In the US, helicopters and other fixed wing craft flown as 'air taxis' account for approximately 18 fatalities a year per every 2.4 million passenger kilometres (1.5 million miles), making them twice as dangerous per passenger mile as cars.

Let's assume all of these problems are solved: batteries are good enough, or manufacturers have opted for hybrids; vertiports (or whatever they end up being called) are in place, as are pilots and air traffic controllers; and regulators have signed off on the whole shebang. Now comes perhaps the biggest challenge: the business model.

Air travel is standard for much of the world, but not inside cities, nor between closely located cities. That's largely because of cost. There's little doubt eVTOLs will be cheaper than a personal private jet, but that's not saying much, given those are hired by billionaires.

There are two main visions, which can loosely be categorised as intracity, or flights within a city, and intercity, flights between cities. These two models aren't mutually exclusive – in theory, we could have both, just as metro services like the London Underground or New York subway carry commuters within a city while long-distance rail services take travellers to different cities. For flights within a city, Uber imagined its on-demand aviation would operate more like a bus than a car: tap a button for a ticket, but then head to your local vertiport to catch your ride. That journey to catch your flying taxi may take longer than you'd like, but the time would be made up in the air, with minutes-long flights beating cars stuck in traffic for hours.

Intercity flights would cover longer distances, perhaps linking up cities and towns that don't have existing rail connections, or are harder to reach because they're on islands or up mountains, for example. Rather than drive for hours or catch a ferry, or travel into a central city and then back out again, cities and towns can connect more directly. It's easy to see this working first where there's money involved, linking a corporate headquarters with a rural outpost, a posh suburb with an office-filled city, or airports with each other or business districts.* The Lilium Jet is geared towards something similar to this approach, starting out life as an electric business jet, or put differently, a flying corporate minibus. Lilium already has a partnership with private jet supplier NetJets for regional flights, so the early eVTOL customers may well be rich dudes getting from Miami airport to their favourite golf course. Wiegand is relaxed about that. 'In the first two years, we will serve that premium helicopter and business jet market,' he says. 'And the reason is it's got the highest margins – there's a high willingness to pay to be the first and that helps us offset the higher cost initially when production volume is small.'

⋆ ⋆ ⋆

None of this sounds like a flying car, does it? There are personal aviation machines in the works that are perhaps closer to that vision, more similar to the roadable aircraft from the 1940s. One is the Model A – you see what they did there, with that nod to Ford's Model T? – from Alef Aeronautics, which is a car made out of mesh metal with the engines in the body to lift the vehicle up to cruising altitude. The entire body then shifts around the passenger bubble to fly

* Lilium's chairman Alex Asseily once suggested to me that the Jet would be perfect to link Heathrow airport to London's business district, Canary Wharf, as there was no direct rail route between the two. While his argument was sound, we had our call on the opening day of the Elizabeth Line (aka Crossrail) that now links the airport directly to Canary Wharf in 45 minutes.

forwards. Looking like something Batman would drive on the weekend, with a slick black sporty exterior, the Model A is actually more like a golf cart – it has a low-speed classification for on roads – and despite no public test flights has won limited flight approval. Unlike some of the 1940s designs, this isn't designed as a car for everyone: you'll have to shell out $300,000 for this roadable aircraft.

The Opener BlackFly is another that is closer to that vision of a roadable aircraft. It's a lightweight vehicle that fits in a trailer and unfolds into a strange all-electric drone that you fly yourself. Though there's automated assistance, you'll need to have flight skills to control the BlackFly, but for an ultralight vehicle a full pilot's licence isn't required in the US. The design is also different from those we discussed earlier. When the BlackFly shifts from take-off to flight, the wings don't shift into position; instead, the entire body of the aircraft tilts. Opener has since rebranded as Pivotal and the BlackFly to Helix, which is selling for $190,000. That price means it's more likely to be used by adventurous execs to travel from their countryside home to work on a Monday morning, or for tourist flights – it would indeed be a bucket-list travel day to cruise one of these through the Grand Canyon, along the Cliffs of Moher or across the Manhattan skyline.

Another company aimed to build a personal, roadable VTOL for anyone to use: Kitty Hawk. Founded by Sebastian Thrun, whom we met making driverless cars, with funding from Google co-founder Larry Page, this secretive startup built more than 100 different aircraft. They included the Flyer, a single-person eVTOL that was classified as an ultralight, so didn't require a pilot's licence to drive.

In 2020 Kitty Hawk shelved Flyer in favour of an electric, autonomous air-taxi model called Cora, as well as a multi-passenger vectored thrust-style jet called Heaviside, which looked more like a rival to Lilium for urban transport. But Thrun couldn't see a viable path to business and Cora too was spun off as Wisk – a joint venture with aviation giant Boeing, before Kitty Hawk itself was itself entirely shut down.

Just like the famed aviators of the last century who had functioning designs and financial backing, Thrun couldn't make personal flying cars feasible. Why? 'Things were going well by aviation standards,' he tells me. But there was too much that hasn't yet been settled: air traffic control, how to make planes communicate with each other, how to build them more cheaply – 'not a million bucks apiece' – and reorganising certification regimes. 'Certify an aircraft today, and you're stuck with the technology for five years,' he says. 'It's been over five years that people have submitted what's called type certification for drones and small aircraft. Those are still in the process of being certified. None of them have accomplished certification with the FAA – and the technology is completely outdated.'

He adds: 'There's just so much to be done: on a political basis; on the development there was more research and development to be done, and it was the right choice in the end to say, look, we can't really build a company yet. It took me a while to realise that.'

Business models have shifted for others as well, including Lilium, which started out making personal flying machines. While studying at the University of Munich, Wiegand wanted to build an electric jet engine, and spent weeks drawing up designs. A roommate pointed out he should not just build it but build a company around it, and so he rounded up a team of founders to cover the key technical areas, including Sebastian Born (aircraft design and structure) and Patrick Nathen (mechanics).

Wiegand didn't have any friends who could design control systems, if you can believe it. So he walked into the engineering department and asked the receptionist if she could round up some soon-to-graduate PhDs in the subject, and inexplicably she did. He delivered a 'not very good' pitch in the kitchen, and though many left mid-presentation, some stayed, including Matthias Meinter.

They set about building a personal flying machine, but quickly realised the technology necessary for it to be autonomous wouldn't be ready in time, and a personal flying car that required a pilot or licence wasn't ever going to make money. For now, they have opted for an OEM (original equipment manufacturer) approach, in which they sell electric business jets to established airlines, in order to start earning money as quickly as possible. 'The initial aircraft was planned to be a sports plane, a two-seater for rich people in an automotive style with high-volume manufacture,' he says. 'And then from there onwards, we wanted to grow the size of the aircraft and make it more suitable for the masses.'

Now, business jets will be how the company enters the market, because they pay, or at least they might. Wiegand wants to make engines that would work in more traditional aircraft – hopefully making aviation more sustainable – but also to go back to the idea of a personal aircraft. Perhaps there's a personal reason: his studies and subsequent work schedule have led to him giving up his pilot's licence as he lacks the time to fly.

Realistically, in the beginning, eVTOLs won't be for most of us. If the wealthy are taking to the skies to dodge traffic, where does that leave the rest of us? Still stuck in traffic. After all, there's little benefit to hopping rooftop to rooftop or flying in from the country house to the city pad for work if we can all do it, as then the roads would be clear to drive.

But the technology powering eVTOLs could offer other benefits in the longer term. Even if never truly autonomous, handing over some of the controls to computers could make piloting much easier, while eVTOLs themselves will be much cheaper than helicopters – not to mention quieter. Consider air ambulances. Right now, those helicopters are expensive to buy, noisy to operate and have limited places where they can land. Because of such challenges, there are only about two dozen in operation in the UK, two of them in London.

Cheaper options could expand rural and urban emergency services, making air ambulances the default for the most serious health emergencies. Indeed, an air ambulance charity in Norfolk, UK has ordered a £350,000 PAL-V Liberty flying car for its emergency services.

So, flying cars aren't going to be parked in driveways anytime soon – if ever. That's good news. Flying cars look fun, but would clog up the skies above cities and exacerbate transport inequities, with the rich flying fast over traffic while the rest sweat it out on buses. But flying electric buses could provide links in transport deserts, where people are forced to drive for lack of other options, and offer an alternative to infrastructure-heavy railways for less population-dense regions. Rather than trundle through national parks in traffic jam-causing minibuses, tourists could quietly glide through river valleys and over mountain peaks, spotting wildlife from above. Air ambulances and police choppers could be cheaper and able to land in more locations with less disruption to the surrounding community.

And, most importantly, short-haul flights will go from emissions-spewing monsters to clean energy – that's not the future promised by *The Jetsons* et al. – it's much better.

CHAPTER SEVEN

Hyperloop

I am standing in the rain in Edinburgh, waiting for a hyperloop. Don't worry, this isn't some sci-fi future-looking scene set 20 or 30 years in the future where I imagine what life will be like when these technologies eventually arrive. (I wouldn't do that to you.) Instead, I am at European Hyperloop Week (EHW) 2023, straining to see through a crowd of hundreds of students wearing team-proclaiming sweatshirts and jackets – merch is a big deal here. In front of us is a short track, and hanging below it like a high-tech bat is a hyperloop pod built by students from Delft University.

The Dutch contingent is one of the larger hyperloop projects, with 40 or so students. They've won previous competitions run by EHW and Elon Musk's SpaceX, and are backed by serious partners like Siemens. Over the microphone, a student describes how the pod works: it oddly levitates downwards, so there's no contact during movement, and features an onboard magnetic motor rather than track-based propulsion. We've been promised the pod will levitate and move forward along the track, before returning, and then run again at its top speed of 5m/s. No wonder everyone at this demonstration day is keen to catch a glimpse.

And that's hard to do. The crowds press against the metal barrier like a pop star is about to saunter by, and as an average-sized woman, I can't see over the shoulders of the much taller Dutch and German lads. One woman sits on her partner's shoulders; children are shoved to the front; two friends try to hoist a third up, but can't hold him long enough. Short, elderly parents are gently scooched closer, their offspring's sturdy peers giving way to the pride on their smiling faces. Cameras and smartphones are held aloft, their owners desperate to get a good clip for social media. I can see what's going on by watching these screens, like an ad hoc network of miniature jumbotrons.

But little is happening, so I'm not missing much. There's a long list of safety and procedural checks to run through before the pod is allowed to go whoosh. The team counts down: three, two, one! A loud click but nothing else. Something has failed. The team restarts and runs through the setup again. We count down: three, two, one! With a thump and a hiss, it fails again. Third time isn't the charm either, with another 'three, two, one!' followed by a clunk and then a disappointed *awww* from the crowd, myself included. We want this to work – I am not a hyperloop supporter, but it's impossible not to be won over by these students with their charming enthusiasm.

Eventually, the Delft team admit defeat, but only long enough to recharge their depleted battery. To be clear: these aren't full-sized pods but tiny prototypes – this is a futuristic model railway show. Student teams work on pods throughout the year and then reveal their progress at competitions like EHW, and previous ones run by SpaceX between 2015 to 2019. There's no mention of Musk, SpaceX or his tunnelling startup The Boring Company here, beyond one attendee's bright yellow shirt proclaiming 'not a BORING competition'. But the projects these students strive all year on are unquestionably tied to Musk, who sparked the hyperloop industry with a 58-page white paper detailing the technology in 2013, before all but dropping out of the race. As yet, no hyperloop has been built beyond a few short test tracks.

What is a hyperloop? It's like flying, but at ground level in a tunnel: the pods are aircraft without wings, they accelerate to 'take-off', and then cruise at high speeds – up to 1,000km/h (620mph) – through low-pressure tunnels, where reduced drag enables more efficient travel.

That idea is hundreds of years in the making. Pneumatic trains have been blowing through investor cash since Victorian Britain, when the era of Railway Mania saw investment in all sorts of wild train ideas before steam rose to the top as the locomotive design of choice. The railway boom was akin to what we've seen with tech cycles, be it Bitcoin or NFTs or generative AI: an idea catches fire and investors empty their pockets in the desperate hope of picking the eventual winner.

Faster, cheaper, sustainable transport is a worthy dream, of course. We shouldn't mock those who can't stop trying, even if it is against all sense, to create something better. With that attitude, we'd never have built railways in the first place. Think through how silly they are: carriages of people and goods hauled by a steam-spewing, coal-burning engine, but only where we lay expensive metal tracks. And that's one of the better ideas.

Eventually the money dried up and steam won out. But, nearly 200 years before Musk uploaded his PDF about hyperloops to the internet, an inventor sat in his London office, sketching out his version of pneumatic transport. Like Musk, George Medhurst wasn't laughed out of polite society – he was widely considered a genius. And like Musk, he didn't build his designs – they were developed by others and brought to life amid the railway craze of the 1800s.

The idea of floating trains to reduce drag didn't live and die amid Railway Mania. Even once railway designs were well established – and we all agreed they involved rails and a locomotive – designers have come up with new versions, from vactrains to hovertrains and even magnetic levitation. These have been tested, trialled and have (mostly) failed over the intervening centuries, meaning hyperloop developers are following in the stumbling footsteps of many who believe there's a better way to build trains, and that it involves sucking or blowing.

But lying beneath Broadway in New York, crumbling in the marshes outside Cambridge in the UK, and sitting unused at railway museums around the world are the rotting remains of these ideas: trains and tracks for hyperloop's pneumatic and hovercraft predecessors that whisper words of caution to anyone who thinks trains should float through tunnels: *no one wants to pay for this.* It's a lesson we don't want to learn.

★ ★ ★

Denmark Street sits around a corner from where busy Charing Cross Road and busier Oxford Street slam into each other in London. This is where George Medhurst worked as a civil engineer, inventing balancing scales and designing an early example of the steam engine. In 1810, he turned his brilliant mind to solving a challenge that remains in modern London: moving stuff around the crowded streets.

That year, he published his solution in a paper with a title that doesn't get exciting until the last word: 'New Method of Conveying Letters and Goods with Great Certainty and Rapidity by Air'. Medhurst wasn't hoping to fling packages above streets, like modern attempts at drone deliveries. Instead, he proposed building tubes underneath the city to send parcels on rail carts propelled through the network by compressed air. He followed up that pamphlet with a selection of alternative designs: first, a pipe laid between rail tracks that pull the carts along the rails; second, a similar scheme on roads to pull what we would now call cars (they were yet to be invented); and third, a tube with a small carriage for goods and a larger carriage for passengers.

Medhurst never built his system. But pneumatic tubes to shuttle packages and messages did come to life a few decades later. Beginning in the 1850s, two entrepreneurial Brits – Thomas Webster Rammell and Josiah Latimer Clark – constructed a version beneath London, selling access to the Post Office so that pneumatic shuttles could carry mailbags. They also carried a few courageous-or-bonkers people (we'll get to that project later).

Pneumatic systems to carry small parcels or letters were also installed across dozens of other cities around the world. The largest covered 450km (280 miles) in Paris and operated until 1984, allowing a message to traverse the French capital in less than two hours. In New York, a pneumatic tube system was carrying 95,000 letters every day by 1897. It was so quick and easy that people could send missives back and forth throughout the day, like an analogue version of text messaging: a husband could alert his wife that the boss was coming to dinner, and she could reply asking about his preferred meal, and receive the reply, all before lunch.

As efficient as such pneumatic message tubes were – and they are so helpful that some remain in use at hospitals and banks today – there were issues. Maintaining city-wide networks was a difficult job requiring creative solutions: Parisians reportedly shot bullets down the tubes, using the ensuing sound waves to locate blockages, while Berliners doused theirs with wine to avoid ice in winter. How very European.* Medhurst's initial plan was for goods, not people, because he didn't think people would want to be 'shot along the tube like pellets down the barrel of an airgun'. Surely passengers would demand windows! But others inspired by his designs ran with the pneumatic proposal and found ways to apply it to trains that allowed for both windows and passengers.

First, there was John Vallance. In 1826, Vallance built a demonstration of an air-propelled, tube-based carriage at his home in Brighton. Around 2.5m (8ft) wide, running on a track 45m (150ft) long, it could carry 20 people, though reportedly much slower than walking speed. To brake, the doors were flung open, releasing the pressure. But it was so jarring for passengers it was dubbed 'Vallance's Suffocation Scheme'. Despite the problems, it did win local support, with a meeting at the Old Ship Tavern in Brighton producing

---

* This is from an interview with Molly Wright Steenson (https://biturl.top/JZV7fu), who wrote an excellent and detailed essay about pneumatic tubes: https://biturl.top/YZVzUz

just a single detractor.* However, plans to build the scheme to link Brighton and Shoreham and eventually London for freight as well as passengers, believing it would be a tourist draw, were scuppered due to lack of funding.

Vallance's Suffocation Scheme never expanded beyond Brighton, but four railways were eventually built using Medhurst's proposal to push and pull rail carriages along using compressed air. Before we get to those, let's consider the railway market at the time. As wild as pneumatic or atmospheric railways may sound now, back then the design and concept of trains as we now know them wasn't considered a done deal. And removing engines from trains and housing them trackside had serious benefits: coal-fuelled steam engines left passengers coughing and covered in soot, and offloading the dead weight of the engine and the coal helped to cut fuel costs. That was particularly important for steep gradients, which were a struggle for heavy early steam engines. This is why cable cars were used in places like San Francisco, as they don't have an onboard engine but are hauled up the hilly streets on a cable that's pulled by a mechanical engine embedded in the ground. When battling gravity, it makes sense to ditch unnecessary weight.

Cable-pulled trains were also tried in Victorian Britain. A cable railway linked Fenchurch Street in central London with Blackwall 4.8km (3 miles) away. Each carriage was attached individually, and could be dropped at any intermediate station if required by passengers, and picked up again upon the train's return. That this was actually built shows how much money was kicking around for anything rail-related – even silly ideas like rollers instead of rails.

---

* The Old Ship remains as a hotel in Brighton, and is moments away from Volk's Electric Railway, the oldest working electric railway in the world, which has been transporting passengers along the sea front since 1883.

So when shipbuilding brothers Jacob and Joseph Samuda and their partner Samuel Clegg built a test track in west London that offloaded a locomotive engine trackside to suck air out of a pipe to push along a piston, it wasn't the oddest notion ever heard.* However, George Stephenson, who built the very first steam engine, called it 'a lot of humbug', because air would leak from the pipe. He wasn't wrong, but as a member of the establishment and creator of a rival technology, he was perhaps predisposed to criticism.

In theory the idea behind atmospheric or pneumatic railways is simple: suck all the air out of a tube in front of a train, and the pressure of the air behind it will propel the train forward. In practice, making it happen is hard, as building a vacuum-sealed tunnel that can fit a train large enough to hold people is no easy feat. Plus, as previously noted, people weren't necessarily keen on being shot down a tube like a Parisian bullet.

The Samudas, being clever, had a solution: push the train indirectly. Building a train-sized, route-length, airtight tunnel is hard. Instead, build a mini version and use that to drag a train along. How? Lay a pipe between the railway tracks and shrink the pneumatic design into that smaller, more manageable space. Inside that pipe there's a piston, just a chunk of metal which is what the air pressure pushes forward. Steam engines situated every few miles trackside suck the air out of the pipe, and the air pressure behind the piston pushes the piston forward like a

---

* They weren't even the first to try the idea. In 1834, Henry Pinkus, an American inventor who had moved to London, wanted to build a pneumatic rail demonstrator along the city's Kensington Canal under the name National Pneumatic Railway Association. While Pinkus also faced problems sealing the pipe valve, using a rope, in addition he couldn't raise the required £200,000 in funds, so the idea was never built, though reports suggest some of the system was built before the site was bought up by another company. These days, it could probably raise the cash in a weekend on a crowdfunding site just by selling T-shirts with the company name emblazoned across the front. I'd buy one.

champagne cork, pulling the train forward via a connecting arm that's attached to the front carriage.

That arm is the inherent weak point of the design, as it requires an opening in the pipe to connect to the piston, and that opening must run the whole length of the vacuum pipe. It should go without saying, but it's pretty hard to maintain a vacuum in anything open. This is where the Samuda brothers' partner, Clegg, stepped up: he designed a continuous valve along the top of the pipe, covering the opening on the top with a flap of leather and metal that could lift when the arm came through before slamming shut again.

The brothers Samuda first built an 800m (½-mile) test line at Wormwood Scrubs in west London, where they leased the same length of rail on which to build their system, advertising that the public could watch every Monday and Thursday as the trains hit top speeds of up to 72km/h (45mph). This proved excellent marketing for the atmospheric propulsion method.*

The system worked, though not without a few problems that required expensive solutions. What would happen if the train had to cross another line? (They solved this by building the first rail overpass.) How would you move an engineless train that wasn't yet connected to a pipe? (With the Croydon line, rails coming into stations were built on an incline and those leaving sloped downwards for a gravity-assisted start; in south Devon, a rope was pulled by an auxiliary vacuum tube that provided the muscle.) How much pressure was necessary to drive a train up a hill or one carrying heavy loads, and how do you account for leakage? (More pressure, so bigger pipes

* After this early work, Clegg moved to Lisbon, Portugal, and Jacob died when a steam engine exploded aboard the *Gipsy Queen* in 1844, so Joseph completed most of the later efforts. Media reports confused the first names of the brothers, reporting Joseph as the one who died. Just a reminder to take all media reports – those more than 150 years old as well as written today – with some caution.

were needed than the ones they used, it turned out.) When do the trackside engines need to be fired up? (The telegraph had just been invented, so real-time communication was only just becoming possible; this meant tight scheduling and often engines were left running, causing waste.) But the biggest challenge of all proved to be the flaps that covered the opening in the pipe. The combination of leather and tallow used to create an airtight seal was delicious to rats, and froze in the cold. Despite some successful solutions, this turned out to be the detail that largely scuppered pneumatic trains for the era.

One example of the technology being used successfully, however, is the Dalkey Atmospheric Railway (DAR) in Ireland, which opened in 1843, linking Dublin and Kingstown to the suburb of Dalkey, south of the city. More of a tourist trap than a commuter line, the DAR was a fashionable and fun day out, with trains running every half-hour and ferrying 4,500 passengers weekly. The steep line was well suited for an atmospheric trial, as locomotives of the era would have struggled up the gradients that the pressurised piston system could handle with ease.

Five minutes before departure from Kingstown, the engine at Dalkey would start pumping, with the hiss of air able to be heard several hundred feet away. Staff then pushed the train into position, lining the piston up into the pipe, with the brakes on to ensure the carriage didn't shoot off on its own. The vacuum then pushed the piston forward, which in turn dragged the carriages. To protect the leather valve cover from the local weather, the Samudas used a long iron flap that was lifted by the wheel of the train as it went down the track, falling back down after the train passed. A heater warmed the tallow mixture (which also included beeswax in this case) to ensure everything stayed soft and sticky and remained in place. Such effort was necessary, as even when the valve was working well, it still leaked.

The DAR averaged speeds approaching 50km/h (30mph) for the three-minute journey. The piston tube didn't actually run the entire length of the trip: the last 457m (500 yards) of the trip was completed entirely from momentum – no easy task when it was the steepest bit of the route. Brake too early, and passengers had to push the train the remaining distance; brake too late, and the train sped through the station and off the rails (this actually happened, luckily without injury). The return was downhill and powered entirely by gravity, so passengers might have to get involved to push it into the station – this was how things rolled on what was considered a successful railway technology.

At least once, the carriage went much faster than it should. As the pumping station started up, the front carriage uncoupled from the rest of the train. Unfettered by the usual weight of the rest of the carriages and the passengers on them, it hurled up the track in a record-breaking time of just over a minute and 20 seconds – which was all a bit of a shock for Frank Elrington, an engineering student who was the only person in the carriage. At an average speed of 135km/h (84mph), he was the fastest man on earth for a good long time. While Elrington survived, the DAR did result in the death of one man, named only as Macdonnell, who was killed while trying to cross the line. Pneumatic systems weren't silent but were much quieter than steam locomotives of the day, so perhaps he didn't hear it before he was 'struck to pieces'.

The Dalkey Atmospheric Railway ran for 10 years, but was replaced by a locomotive after the engines powering the pneumatic operation broke, and local train operators decided to link their lines up on a single gauge rather than force travellers to change carriages multiple times. To keep the line running, a locomotive was brought in – the *Princess*, the first built in Ireland – and reconfigured so the smoke stacks fit under the low bridges. At this point, it was clear that steam could work in the region, despite previous concerns about the steep gradient.

Still, during the first few years of operation the Dalkey Atmospheric Railway acted as excellent marketing for the Samudas. The French government sent representatives to understand the system, eventually building an 8km (5-mile)-long atmospheric railway at Saint-Germain. That line was closed in 1860 due to a fatal crash that killed three including the driver in 1858, and growing preference for what is now standard rail. And in 1844, the already famous engineer Isambard Kingdom Brunel showed up in Ireland with colleagues from South Devon Railway. Inspired, they went on to build a 30km (20-mile)-long rail line running along the sea from Exeter St David's to Newton Abbot using atmospheric propulsion. The first section opened in September 1847 with speeds up to 110km/h (70mph), but a year later, the entire effort was shut down because the valves failed, Brunel having opted against using the Ireland-developed iron plates for weather protection.*

At the same time the other railways were being created, Samuda had his own go at building a pneumatic piston railway through the south of London, backed by another famous engineer of the era, Sir William Cubitt. The aim was to link London Bridge to Croydon, with an eventual extension down to Epsom, Surrey, 16km (10 miles) south of the capital. The piston-driven line ran alongside the steam railways headed for Greenwich before turning south.

Work started on the first leg linking London and Croydon in October 1844, and only a year later, the first trial run was held, with 10 carriages filled with curious, courageous passengers. The journey took just under seven minutes to travel the 8km (5 miles), topping out at speeds of 84km/h (52mph). However, that was the high point of the project and technical problems soon arose. The trial section used 40cm (15in) pipe to house the piston, which didn't generate enough pressure to drive the trains. The lack of a telegraph meant schedules had to be carefully kept or engines weren't ready to suck out air or did it too early, costing fuel. And, as before, weather and rats

* You can check out one of the few remaining sections of pipe at the Didcot Railway Museum in the UK. Of course, the leather valve covers are long gone, presumably in the belly of a long-deceased rat.

degraded the valve flaps, with so many dead rats piled up in the pipe that their corpses clogged it. The costs to fix the valve so terrified the company directors that the atmospheric design was abandoned in 1847 in favour of steam locomotives, and the line continues to this day, though running electrically.

The failure of the Samuda brothers* to make their piston-based pneumatic railways work didn't end the belief that atmospheric railways were the future of transport. Earlier I mentioned Thomas Webster Rammell and Josiah Latimer Clark, who patented a way to send messages and packages via pneumatic tubes. Latimer Clark began the efforts in 1853 with a 4cm (1½in)-wide tube that blew telegraph messages from the local telegraph company to traders at the London Stock Exchange just a couple of hundred metres away. A few years later, Clark teamed up with Rammel to co-found the London Pneumatic Despatch Railway (LPDR),† filing a patent in 1860 for an 'improved pneumatic railway' that could work at any size for goods or passengers. They began in Battersea, London in 1861 with a 76cm (30in)-wide cast-iron test tunnel along the southern edge of the Thames. Just 413m (452 yards) long, this didn't use a piston in a pipe; instead, a steam engine was used to pressurise the entire tunnel to push forward wheeled carts that created a seal with the tube walls using vulcanised rubber flaps – the very material that might have made the Samudas' schemes viable was now widely available. Speeds topped 64km/h (40mph), and local thrill-seekers loved to ride it for fun.

---

*Joseph Samuda died in 1885. In the intervening years, he went back to shipbuilding, volunteered in the military and became a Liberal member of parliament. He's buried in Kensal Green cemetery. One of the ships built at Samuda's Cubitt Yard works was the HMS *Tamar*. Launched in 1863, it remained in use until 1941 when it was scuttled in Hong Kong harbour to prevent it from being taken by Japanese forces.

† It was originally named the Pneumatic Parcel Post Company, but that was deemed too similar to (and suggestive of a connection with) the General Post Office, so it was renamed.

The carts were tiny and fit snugly inside the 1m (3ft)-high pipe, which was shaped like a miniature train tunnel, with flat sides and a curved ceiling. The small size required passengers to lie flat or crouch down to fit. It would have been like riding a very dark rollercoaster without any safety features at all. No one appears to have died, though one newspaper report describes the cart being 'in a shape not unlike a coffin', saying that with just a few inches between your face and the tube, 'if, unfortunately anything happened to retard the progress of the vehicle, there was a very disagreeable prospect of being entombed alive in an iron vault.'

The LPDR considered that enough of a success to build a short line between Euston and a local post office on Eversholt street, just around the corner. Opening in February 1863, it could transport mailbags from the railway to the sorting office in just a minute, the operators charging the Post Office for this faster service. Further lines were built over the next several years, linking Euston station to Holborn and then on to St Paul's and finally the General Post Office. The carts were now reaching speeds of 100km/h (60mph).

However, despite the speed, the system wasn't efficient at transporting mail. The damp tunnels left mail damaged, and little time was actually saved as the network was underground so the sacks of letters and parcels had to be ported downstairs at one end and then back upstairs at the other. Beyond that, it was deemed too expensive by the Post Office, the one potential customer. But for those several years in the 1870s, there was a tiny, miniature rival for the then-nascent London Underground.*

It was clear that people were inexplicably willing to ride in an air-propelled train, even if it wasn't the most efficient system

* In the end, an underground mail network was built using more standard rails; called the Mail Rail, it only closed in 2003, and you can ride a section of it at London's Postal Museum, which also has one of the pneumatic rail cars on display.

for shuffling mailbags. So in 1864, Rammell spent eight months building a short test track for a passenger-carrying atmospheric railway in Crystal Palace Park in south London. It took 50 seconds to cover the half-kilometre track mostly set inside a bespoke tunnel. As with the LPDR system, there was no piston pipe involved in this railway: the carriage itself was what was pushed along. The line was built in a brick tunnel around 3m (10ft) wide, with the 35-seater carriage pushed forward via a massive 6.7m (22ft)-wide fan powered by a disused steam locomotive. Air pressure was partially maintained by a large ring of bristles attached to the rear of the train to create something of a seal along the tunnel walls. On the return journey, the fan created a vacuum to pull the carriage back; it's unclear whether passengers could disembark after the first half of the journey, or even if there were stations at both ends.

The system ran as a tourist attraction costing sixpence per ride, and was described as 'luxurious' by the same journalist who raised the concern of the mail cart becoming his iron coffin in Battersea. It was also a proof of concept designed to sell the technology. Some welcomed the air-driven train, with *Mechanics Magazine* predicting that with this technology as a rival, 'the locomotive will be unknown on underground lines'. But praise was not unanimous. The *Engineer* described the stationary engine as 'like an imprisoned racehorse' and said the two entrepreneurs 'now desire not only to pump and blow her Majesty's mails, but her subjects generally through exhausted and flatulent tubes'.*

---

* It wasn't only technical papers that offered criticism: Marian Evans, better known as George Eliot, visited the Crystal Palace railway, including a reference to pneumatic systems in her novel *Felix Holt*: 'Posterity may be shot, like a bullet through a tube, by atmospheric pressure from Winchester to Newcastle: that is a fine result to have among our hopes; but the slow old-fashioned way of getting from one end of the country to the other is a better thing to have in the memory. The tube-journey can never lend much to picture and narrative; it is as barren as an exclamatory O!'

The railway was dismantled after two months. Rammell planned to use the same system to link Waterloo on the south bank of the Thames with Whitehall on the north the next year, and even dreamed of a line under the English Channel to Calais in France and a route through the Alps. While parts of the Waterloo–Whitehall line were built, a wider financial crisis meant funds dried up and the work was never finished. Macroeconomics took out these projects just as readily as the rats.

* * *

The idea then emigrated to New York. Inspired by the British efforts after a visit to Crystal Palace, entrepreneur Alfred Beach built a pneumatic people mover, demonstrating in 1867 at the American Institute Fair a 1.8m (6ft)-wide model of a pneumatic subway system that used air to push the carriages, with a giant fan at one end. That prompted further trials, and his Beach Pneumatic Transit Company built a 90m (30ft)-wide test track under Broadway between Warren and Murray Streets, though the eventual aim was to run it 8km (5 miles) to Central Park.

Beach was given permission for the work on the grounds he was building a package-delivery tunnel, but instead he built a test track to carry passengers. It took just 58 days to dig, helped by a tunnelling shield of Beach's own design. The very first day it opened, 26 February 1870, the engine failed, but after it was repaired a week later, visitors entered at street level via a clothing store to visit the tunnel, enjoy the lavish platform and hop a ride, with the 25-cent fares donated to a charity for war orphans.

Over the course of a year, more than 400,000 rides were taken on the short track, with the carriage only able to seat 22 passengers each trip. Even the fan, dubbed the 'Western Tornado', was built to impress, with its 'prettily frescoed wood-work'. Though there were critics, the press was positive. A report in the *New York Times* said visitors 'came away surprised and gratified'. The prestigious journal *Scientific American* was even more enthused by the pneumatic tunnel, saying it was 'important work', and that detractors simply didn't understand 'the benefits which its successful completion are destined to confer upon the city itself'. Among these benefits, the journalist compared travelling above ground in the 'filthy, health-destroying, patience-killing street dust' to the tunnel, which was 'comfortable and cosy as the front basement dining-room of a first-class city residence'.

He goes on, espousing the benefits over steam travel, including the fact that the compressed gas won't mess up your clothes like coal-driven engines. It was also fast, cheap and could very well be the transport method to solve all New York's woes, but also connect the city to points further afield. In short, *Scientific American* really loved the Beach tunnel. And it's no wonder, since, as the article admittedly notes, Beach was a co-founder of the magazine – tech founders controlling the media is nothing new.

The line, however, was never extended. A planning permission battle ensued, with Beach pinning the delays on then mayor William 'Boss' Tweed, though there was also opposition from locals who preferred the idea of an elevated train network, fearing an underground railway would destabilise buildings. While approval was eventually granted, a subsequent stock market crash took the wind out of Beach's underground sails, so we simply don't know if his railway would have worked with further development, but in 1900 the point became moot as construction of the existing subway network began, scuppering the project permanently.

Had Vallance, Samuda, Brunel or Beach dodged all the bad luck, economic uncertainty and rats, could pneumatic railways have succeeded? It's impossible to say without the lines having been fully built, but if you chuck enough cash at anything, you can buy your way out of practical and engineering challenges. Projects fail when you run out of cash before you run out of problems.

But the idea of low-friction train journeys holds such promise that it continues to pop back up time and time again, though in slightly different forms. Robert Goddard described his 'vactrain' via an article in *Scientific American*, and after his death, his wife Esther was awarded a patent for the design. In 1965, an engineer for Lockheed Missiles laid out his plans for a pneumatic tube linking Boston to Washington, DC in just 90 minutes, with the tubes themselves floated in water for a smoother ride. Robert Salter, of the military-affiliated RAND Corporation, published designs for high-speed transit using 'evacuated tubes'. Pneumatic, atmospheric trains just won't go away, despite the many historical failures. But you don't necessarily need a big tube to make a railway that runs on air – just look at hovercrafts.

⋆ ⋆ ⋆

In a boggy bit of rural Cambridgeshire in 1973, Research Test Vehicle 31 is whooshing across the misty landscape at 160km/h (100mph), powered by magnets and riding on a cushion of air. The team celebrate the record-breaking test but it is also a commiseration: that very same day, the entire project has been officially shut down for good.*

It's a century on from the Samuda brothers and Beach. Rail is well established, but cars are the order of the day. At the time, the British rail network was considered one of the finest in the world, according to the *New York Times*. British Rail has been keen for trains to run faster to better compete with road and air travel, and the government has been aware of new rival railway technologies being investigated

* Now, RTV31 is parked at a trackside attraction-cum-nature park called Railworld, where the only whooshing is passengers on the adjacent line zooming through Peterborough, England.

elsewhere – Japan's Shinkansen were launched in 1964, with the aim to capitalise on recent developments by a local academic who had built a new type of motor. So in 1967, the British government set up a company, Tracked Hovercraft, to explore the idea of floating trains for hyperfast speeds.

Hovertrains looked to use the same principles as hovercraft, floating inches above a fixed track rather than being pulled forward on rails. The train was shaped like a small letter 'n', with the track a concrete beam running through – picture a monorail and that's essentially it. The train is floating, pushed up into the air by fans, though magnets were also investigated. That air gap means there's less friction, so less power is required to move it forward. Indeed, various reports on hovertrains show they could be pushed by hand once floated. Because of this, early prototypes used a propeller or aircraft engine for propulsion, meaning these were essentially grounded aeroplanes.

But the Tracked Hovercraft design took advantage of a recent development by a British academic, Eric Laithwaite.* A professor at Imperial College, London, he took advantage of magnetic repulsion – when you try to force two magnets together and they push away from each other – to build a levitation system dubbed the 'magnetic river'. The idea was to use the repulsive force to 'float' trains. Unfortunately this isn't stable; the force sends whatever you're floating off to one side. To solve this problem, Laithwaite built an electromagnet – a

* While now he's known as the 'father of maglev', in the mid-1960s Laithwaite was already a household name (well, depending on the sort of household), appearing on television and delivering 1967's Royal Institution Christmas Lectures, a high-profile annual tradition. Amidst all this work, he was also under consideration to be the scientific advisor on the famously long-running British classic TV series *Doctor Who.* Laithwaite lost out on the gig to ophthalmologist Kit Pedler, but still wrote and submitted a script for the show, called 'A New Dalek Adventure', in which an atom-rearranging creature mimics the Daleks. The script wasn't made into an episode, sadly.

magnetic field created by running current through wires around a metal core – that layered multiple magnets so the field pushed inwards rather than out, giving stability to whatever is floating. In a presentation filmed in 1975, he shows a piece of aluminium floating in the air over the magnet; it can be easily pushed forward by his finger or by a small propeller stuck on the top.

But Laithwaite didn't stop there. He combined the idea of magnets with a linear motor. A standard motor spins, but a linear motor is sort of flattened out to produce motion in one direction, in a straight line. By arranging the magnets in rows and running current in the right way, he created a travelling magnetic field, moving forward whatever was floating. This linear induction motor (LIM) could lift and move a train using the same electromagnets. He compares the system to chucking a piece of wood into a river: it floats and moves downstream, hence the name 'magnetic river'.

If all that sounds a bit confusing, don't panic: as Laithwaite admits, we still don't even really understand how magnets work. The short version is this: when arranged in the correct way, not only can electric-powered magnets lift a train up, but they also move it forward.

The Tracked Hovercraft prototype was built in 1963 and a model using LIM for propulsion in 1966. And then they decided to build a life-sized version, picking a flat but boggy bit of rural Cambridgeshire for the venture. In 1969, 1.6km (1 mile) of the system was built, with giant concrete pylons rising up through the Fens between two rivers north of Earith – the 22m (75ft)-long concrete beams were so massive that a local pub was partially knocked down to make room for the lorries.

The very first public test was a slow 19km/h (12mph) and it took another two years to regularly hit near 160km/h (100mph). The first time the big fan was tested was at Tracked Hovercraft's offices in Ditton Walk, Cambridge. The team was planning on simply flipping it on during the day, but realised that the heavy power draw might impact the electrical grid. Being community-minded folks, they waited until 9pm. 'Unfortunately we did take out quite a bit of Cambridge when we did,' says Tony Claydon, who worked on the project. 'We were pleased we'd done it at night, not during the daytime; that would have been very unpopular.'

By all accounts, it was a fun place to work. Claydon looks back on that time as the highlight of his career, while Joan Lobban, who worked in accounts, recalls bosses mingling with staff and sending out for ice cream on hot days. But building the track was difficult owing to the terrible winter weather. 'After good progress, it was a terrible blow when the huge footings and columns collapsed, causing great expense and setback to the whole project,' she tells me. 'This was the first time I saw complete dejection, lack of spirit and an air of depression all over the buildings – both at Earith and Cambridge.' She adds that it was also tough on personnel who had gone overseas to Japan and Canada to scope out rival projects.

There were other problems, too: the magnetic hovering system was heavy; the airflow system chewed through a huge amount of energy; and to make it all work, the power supply was eventually moved onto the train, adding to the weight when the whole point had been to offload as much as possible. Plus, it was incredibly noisy.

In February 1973, another public test was run – and even aired on the BBC. It topped 167km/h (104mph), despite facing heavy headwinds. The relevant minister at the time, Michael Heseltine, said no decision had been made to cancel the project, yet two weeks later it was shut down. Knowing that the project was very likely set to be cancelled, as there was a 'special meeting down in London' that a pair of Tracked Hovercraft employees attended, the team headed to the track for one last go, hoping to once again top 160km/h (100mph), Claydon says. 'I think that somebody actually kept their hands on the power supply so it didn't drop out to make sure we did in fact get there. We wanted to prove that we could do it.'

The project was shelved, and Claydon found out on the radio himself — the Tracked Hovercraft employees couldn't get to a phone to call the team, as journalists hogged them in efforts to file their stories.

We can't answer why the government was so callous in shutting down the project,* but there were good reasons to put investment elsewhere. After three years, it hadn't yet delivered on its 480km/h (300mph) promise, though Claydon argues that would have been possible had a longer track been built; the short distance meant the train couldn't hit top speeds before it had to brake. Elsewhere, efforts to develop the maglev idea showed there may be better techniques.

Alongside the Tracked Hovercraft project, British Rail's research lab in Derby had been running the Advanced Passenger Train (APT), through which researchers were investigating new high-speed techniques that would work with existing rail, such as tilting, reducing drag and other tweaks.† The APT project was expected to be ready more quickly than Tracked Hovercraft and wouldn't require a

---

* Claydon believes the government just lacked imagination, not helped by the fact that visiting the Tracked Hovercraft site was less exciting than rival transport technologies, as the train couldn't take passengers, meaning officials stood in a temporary shelter and watched the train go by – hardly an impressive demonstration of the future of transport. Had the project been wound down more carefully, Claydon says, the technology they developed might have been put to good use. Or, as the report put it: 'They thought that they were closing down a project with dubious commercial prospects when in fact they were closing down the one major centre in this country of a new technology centre.' One struggling idea doesn't mean you shutter the whole lab, after all.

† Early development of the APT-E (E for 'experimental') proved controversial among train driver unions because there was only one seat in the cab. Union ASLEF 'blacked' the project, effectively banning any members from working on the project. But, notes Professor A. H. Hickens, a locomotive inspector helpfully moved a train one night so that modifications could be made. That sparked a day-long strike which cost more than the entire annual R&D budget for the APT-E project.

massive infrastructure rollout; all those concrete pylons come with a cost.

Rather than a full-scale reimagining of how trains could operate, APT tinkered around the edges, experimenting with gas power, a new lightweight, aerodynamic shape, and the ability to lean into curves for faster speeds. While development wasn't cheap – the government spent at least twice as much making APT as Tracked Hovercraft was granted, it would allow speeds up to 240km/h (150mph) by 1980 without new rail being laid, whereas rolling out the hovercraft infrastructure would cost £250,000 per mile.

However, others saw value in the Tracked Hovercraft project. Like the British, the French wanted high-speed rail. Enter Jean Bertin and Aérotrain. Aérotrain worked similarly to the Tracked Hovercraft, in that it required a T-shaped concrete guideway, above which it hovered. However, Bertin didn't have the benefit of Laithwaite's LIMs for propulsion; but he did have a creative solution revealing his aerospace training: propellers. In 1964, for the first tests of a half-scale model down a 3km (1.8-mile) track, the carriage, a sleek, modern, metal contraption with wide windows at the front, no wheels and a propeller plonked on top, looked like a tastefully misassembled plane.

That design couldn't go fast enough so the propeller was eventually replaced by rockets, making the Aérotrain look even more like an aeroplane than before and boosting speeds from 200km/h (120mph) to more than 300km/h (180mph). The team graduated to full-sized prototypes, which in the early 1970s topped 430km/h (270mph). Two significant test tracks were also built near Gometz-le-Châtel, where one section of the track still sits as a memorial to the project at the centre of a roundabout. A third track stretched 18km (11 miles) north of Orléans and now sits abandoned.

Actually, the Aérotrain topped 450km/h (280mph), according to Daniel Ermisse, who was employed as a test pilot on the project. He'd just started working as a television repair man when his uncle, who had worked on construction of the track, introduced him to the Aérotrain project. The lead pilot, Maurice Lefrant, needed a second pilot to help carry out testing. Ermisse drove all of the trains, saying the acceleration was remarkable. 'My speed record on the i80 Aérotrain was 450km/h,' he told me via email, 20km/h faster than the official record.

Despite this development, instead of Aérotrains, France is crisscrossed by TGV, a more conventional high-speed network. Why? Once again, the inability to use existing infrastructure proved a hurdle too high for government officials and transport authorities.

★ ★ ★

That wasn't the end of hovertrains. As with atmospheric railways, Americans saw what was failing in Europe and tried to make it work at home – indeed, California-based Rohr Industries even based its design on Bertin's, with a prototype sent to the government's transportation technology centre in Colorado.*

Hovertrains make sense in the US for the same reason they're a problem: it's a big country. Faster speeds would allow trains to compete with air travel, but large distances also mean high infrastructure costs. The Rohr design, for example, required an electricity substation every 8km (5 miles). Government funding was freed up to develop high-speed trains, sparking the production of prototypes researchers dubbed Tracked Air Cushion Vehicles (TACV), with a test track built in Colorado. Generally, the idea was to build a Bertin-style hovertrain powered by LIM, mixing and matching different aspects of technology.

---

* The local Pueblo Railway Museum has been the benefactor of this test track, with a wide and wild selection of so-called 'rocket cars'.

The TACV programme started with Garrett AiResearch building a LIM-powered, wheeled train that ran on normal rails. The Linear Induction Motor Research Vehicle (LIMRV) topped 300km/h (188mph), and the addition of jet engines helped it reach 412km/h (255mph). Because it went so fast, the LIMRV needed to be slowed down before braking, with a hook on the back grabbing a cable linked to anchor chain at the side of the tracks. Once it slowed enough, the brakes could be applied. Two subsequent prototypes were built, the Grumman-designed Tracked Air Cushion Research Vehicle (TACRV)* and the Rohr Industries Urban Tracked Air Cushion Vehicle, hitting speeds of 145km/h (90mph) and 233km/h (145mph) respectively.

The triumvirate of prototypes faced similar problems to their French and English counterparts, however, including noisy operations and huge energy demands, but what really scuppered the work was that the funding ran out. Hovertrains may not have been able to top 480km/h (300mph) as promised, but their designers certainly blew through a lot of cash.

Hovertrains of a sort do exist, they're just maglevs now. These make use of Laithwaite's magnetic river to float and drive forward a train. Plenty of work has been put into getting them to work, but examples still remain few and far between.

* Ron Roach of the Pueblo Railway Museum in Colorado notes in a video on C-Span (https://biturl.top/6BZfaa) that the Grumman was used to test aerodynamics for the space shuttle. How does he know this? First, when the TACRV was still in a local aircraft museum but set to be moved to its new home, he called up 'a guy in Washington, DC' who asked how Roach even knew about the machine at all. He replied: 'I'm looking at it.' The official was aghast, believing Roach had broken into a government warehouse somewhere, before Roach set him straight. The TACRV was then declassified for museum use, in the event revealing that its shape was in fact designed to look like a space shuttle in order to test aerodynamics. Looking at it, it certainly does share the shuttle's nose.

Think of maglev, and you may well be picturing Asian trains – the only trains currently using the technology are in Korea, Japan and China, so you wouldn't be wrong. But the very first commercial maglev, opened a full decade before other commercial lines, was in 1984 in Birmingham, UK, linking a railway station and the local airport. Sadly, it shut a decade later due to costs – it's a lot of new technology to get people and their suitcases a short way. Germany had a test driverless maglev in 1984 called the M-Bahn, for 'magnet train'; it ran for two years but was closed in 1991 after reunification led to changes in the network. A further Transrapid maglev test track in Lathen, Germany was scuppered after a 2006 crash killed 23 people, though the tragedy had nothing to do with the technology, but human error after a maintenance car was left on the track.

Today, there are six maglev lines in operation. Two are in South Korea: a short 1km (0.4-mile) line at the 1993 Taejŏn Expo site, and a 6.3km (4-mile) link to Incheon airport, with operations starting in 2016. Japan has a 9km (5.6-mile) line running at speeds of up to 100km/h (62mph), also built for an expo. China has three maglev lines: a short metro line in Beijing that tops out at 110km/h (68mph); another linking Changsha with the local airport at 140km/h (86mph); and then one in Shanghai, which uses Transrapid technology and can hit speeds of 431km/h (268mph) – which is frankly ridiculous, as it is a short link between the city and an airport, so that maximum speed is held for just 50 seconds before the brakes go on.

Maglev technology clearly isn't easy, given the long gestation period. But it does at least exist now – which is more than can be said for hyperloop. It takes decades to prove and perfect a new train technology. Maglev may simply be ahead of the curve, or perhaps hyperloop will eventually arrive after several more decades of work. Fast trains are slow in coming.

⋆ ⋆ ⋆

Back in Edinburgh, we wait for the Delft team to recharge the hyperloop pod for another attempt at a demonstration. In the meantime, there's a demo from ETH Zurich's SwissLoop, so the crowds rush to the neighbouring car park to get a good view of the team's 'Piccard' pod zipping down the short track. To further impress the judges, the pod is levitated and pushed back and forth by hand along the track between two team members in hard hats and hi-vis vests, like a costly but slow game of air hockey.

An earlier demo, from Germany's Mu-Zero Hyperloop, also focused on levitation, with its pod sans chassis floating in a metal frame. The smoothness with which it levitated impressed those standing around me – as did the pod's payload capabilities; it kept floating despite several 12.5kg (28lb) weights being added to a box atop it, until it managed 100kg (220lb) while still hovering half a centimetre in the air. After a few procedural checks, the pod lowered back down, softly resting onto the metal as though the 100kg wasn't there.

The science works, in other words. But as impressive as these engineering student projects are, there's not yet a hyperloop in operation, though to be fair, they only really got their start in this form back in 2013 when Musk dropped his 'Hyperloop Alpha' white paper.

What is a hyperloop? There are three main elements, though, of course, the finer details are where the problems happen. The first element is a low-pressure tube – it doesn't need to be a perfect vacuum but the more air that's removed, the less drag there is, meaning less energy is required. The second element is a levitation system, such as using magnets or compressed air; this keeps the train hovering inside the tube for a near-frictionless ride. Finally, the third element is that there needs to be a way to get the pods moving through the tube and, depending on the design, to give them a bit of a boost now and then.

Musk's proposal begins with an opinion column against high-speed rail, or at least a local Californian project, complaining that the home of Silicon Valley shouldn't have transport as slow and expensive as the proposed bullet train. Instead, Musk calls for a 'truly new mode of transport' that he describes as the 'fifth mode after planes, trains, cars and boats'. That would be a hyperloop, at least for situations where two cities with high traffic are less than 1,500km (900 miles) apart – any further, and we should all just fly. Musk wrote that, short of inventing teleportation, the only option for superfast transport beyond air travel is to build a network of tubes above or under ground.

One extreme, he writes, is to build an updated and enlarged version of the pneumatic tubes that used to send mail and packages around cities, with massive fans pushing people-sized pods from LA to San Francisco. The other extreme is to build a massive tunnel with a total vacuum and train driven via electromagnetic suspension – the vactrains described by RAND – but one leak and that system fails. Musk's alternative version sits between the two, with a low-pressure system using pumps to overcome any air leaks.

In the years since Musk's hyperloop paper, designs have evolved, and details have been filled in. Generally, a hyperloop network requires a concrete or similar tube around 3.5–4m (12–13ft) in diameter to be held above ground level on pillars, though some subsequent designs have suggested digging tunnels underground. Hyperloops require mostly straight journeys, with any turns as wide as possible to avoid slowing down. Earthquakes and other ground fluctuations could damage the tubes (just like trains with rails) so raising them up on pillars means movement-dampening equipment can be included. Most plans have a tube running in each direction.

Inside the tubes, the air pressure is lower than outside, so they need to be sealed. The low-pressure system is maintained using vacuum pumps that need to be placed every now and then along the route; exactly how this is arranged depends on the design. Another place of differentiation is levitation systems. Some of the systems use maglev, which requires superconducting magnets. The magnets are fixed onto the pod with metal loops along the track or guideway. Depending on the orientation of the magnets and loops, they repel or attract each other. This gives them three jobs: first, they levitate the pod (or train, as this is how maglev trains work, too); second, they balance it horizontally for stability. And third, metal loops have a current running through them to create magnetic fields that push and pull the pod along. This requires electricity to be run along the entire track, but allows for a super-fast, smooth ride.

The pods aren't like trains. Most current designs are for single pods with space for 40 or 50 people in airline-style seats. Made out of materials like carbon fibre, the pods are generally windowless, as there's no view out anyway – sorry George Medhurst – and embedded with sensors to ensure as smooth as possible running.

Now, Musk in his PDF specifically argued for a raised tube on pylons running alongside the Interstate 5 Highway in California; by including dampers, the system could mitigate against earthquakes, a challenge in the local area. On costs, Musk believed it would be cheaper than the planned rail. The trip described in his proposal, between LA and San Francisco, would be 35 minutes long, with pods leaving every two minutes, each carrying 28 people. For a tube in each direction and 40 pods, the back-of-the-napkin budget puts the cost at $6 billion and suggests one-way tickets could be priced at about $20 plus operating costs.

Sounds sensible, wonderful even. So why didn't *he* build it? Musk has offered various excuses throughout the years: he simply didn't have time between Tesla and SpaceX. He regretted even mentioning the whole thing, claiming he always intended it to be 'open sourced'. There are a few other possible reasons: some say the technology sketched out in detail in that PDF wouldn't work; the costs aren't accurate and would be far higher than budgeted; no government would invest billions on an untested transport technology in a region desperate for mass transit; and so on.

There's another theory, suggested by Musk's biographer Ashlee Vance: Musk never meant to build it, but suggested it to get California's government to delay or cancel the high-speed rail plans.

That said, Musk has revived the idea throughout the years, pledging to build a hyperloop running through the northeast of the US, saying he had 'government approval' for a tunnel linking New York and Washington, DC. But in 2022, SpaceX quietly removed a test tube from its own parking lot, reportedly on the order of city authorities, after it had blocked pedestrian access for six years. The whole shebang however did give us a new Musk-founded company for building tunnels quickly and cheaply: The Boring Company – worth it for the name alone.

While Musk may or may not be serious about the hyperloops, his PDF sparked an entire industry, including companies like Hardt, Zeleros, Hyperloop Transportation Technologies (which recently abandoned an attempt to go public) and Hyperloop One. The latter has shifted away from passengers to freight, which raises the question: what boxes need to be moved that quickly at such a high cost?

Here's a brief history of some of the hyperloops that have been announced with fanfare only to go on and quietly die or fade into nothingness. Hyperloop TT built a test track in Toulouse, France; though it was marketed as the world's first and only full-scale test system, the track was just 320m (1,050ft). It's since been shuttered with a new track planned for Italy. The company also announced projects that slowly faded into nothingness in Slovakia and China, and completed a $1.2 million feasibility study in 2020 for a Great Lakes Hyperloop linking Chicago, Cleveland and Pittsburgh, but that's remained on paper. Hyperloop TT is keeping the dream alive with a more recent contract for a feasibility study in Veneto, Italy, an in-development freight system dubbed HyperPort, and deals with serious rail companies including Hitachi.

Spanish-based Zeleros is working on a port project first, as is Netherlands-based Hardt, though the latter also has a 30m (100ft)-long test track – yes, that short – and has ambitions to build a network linking Amsterdam's Schiphol airport with others across Europe.

Virgin Hyperloop One, once but no longer backed by rich Brit Richard Branson, built a test track in 2020, sending actual humans (both riders were employees) down it at speeds of 172km/h (106mph) for half a kilometre; though it's the longest hyperloop test track in the world, the short distance means the test lasted less than seven seconds. The company has touted routes in Saudi Arabia, but has since pivoted to freight only, laid off staff, and is now apparently working on developing the world's first hyperloop to shuffle packages across the UAE.*

My personal favourite not-yet-a-hyperloop is a line proposed by TransPod, running between my hometown of Calgary, Canada and the provincial capital city of Edmonton, a three-hour drive away. If ever there was a place ripe for a rail line linking cities, this would be it, but that region of Canada is highly dependent on the oil industry and as such obsessively loves cars. Various plans to run a traditional rail line between the two cities have been mooted, including via government and think-tank feasibility studies in 1985, 2004 and 2008. Each of those recommended installing high-speed rail links, although a 2014 report said it was a bad idea due to low population numbers — Calgary and Edmonton each have around a million residents, more than plenty of cities linked by rail elsewhere. Either way, nothing seems likely to happen. Why install train infrastructure when people can just

---

* The entirety of the time I was writing this book, its website was in construction mode; it didn't have a little animation of a builder, but the classy equivalent – a message saying 'our new website is coming soon'. If you can't build a website within several months, I'm not getting in your 700km/h (435mph) pod.

drive? Those without a car or driver's licence can take the bus. A nice, oil-burning bus.

But Transpod believes it can – and will – be done. We know this because of a feasibility report. This is a document written by Transpod about whether or not Transpod's own technology would be feasible; it's marking your own homework. And this marketing document/feasibility report offers the following sales case: the Calgary–Edmonton route is ideal because 14 million car trips are made between the cities each year, with a further 4.2 million journeys between either city and Red Deer, which lies equidistant between the two. Transpod expects a one-way ticket for a journey that would take an hour to cost CAN$90, which is remarkably reasonable given the bus costs as much as $77 and takes four hours.

Transpod predicts that building the line would cost CAN$22.4 billion while the stations and other infrastructure would cost $6.7 billion, plus roughly $1 billion in annual operating costs. But alongside the revenue from a lot of $90 tickets, the system is designed to also carry freight. The combined revenue means the whole thing won't require public funding (they magically never do) and it'll break even in 20 years.

This feasibility study claims that the system would operate like an urban subway rather than an airline: no pre-booking of tickets, just show up at the platform, wait for the next train and board. Each pod will carry 50 passengers, leaving every couple of minutes – a claim frequently made by other hyperloop companies. If TransPod managed to run trains every two minutes, it would be one of the most frequent services in the world. I know this because I have the good fortune to live along the route of the London Underground's Victoria line, which is the second most frequent metro service on the planet, pipped only by the Moscow Metro. It's not impossible, but there's a difference between possible and likely.

Plenty of headlines scream that TransPod has found half a billion dollars in investment, suggesting the project is merely awaiting approval from the local government to start slapping its tube together. But head over to the local government major projects website and there's a listing for a rival project called the Prairie Link Rail Partnership that has a memorandum of understanding with the government to start working on pre-development and consultation and approvals – you know, the boring paperwork side of things. This is a plan by engineering firms AECOM and EllisDon to build an electric or hydrogen fuel cell-powered high-speed train that will link the cities and smaller towns along the way in just 90 minutes. Oh, and it'll only cost CAN$9 billion, which – as ever – won't require public funds. Neither the hyperloop nor high-speed train is likely to happen, but if you were making the choice, which would you go with?

Hyperloop may or may not ever be technically feasible, though it's currently an excellent research project for students and gives plenty of work to transport consultants. But governments need to pick technologies that they know work, and for hyperloop startups there's one very serious challenger: trains.

Existing trains are fast. Italian Frecciarossa zip along at 300km/h (185mph). Japanese Shinkansen speed between cities at 320km/h (200mph). If and when the UK's HS2 exists, it's promised to top 360km/h (220mph). All of this suggests a top-end high-speed train could link San Francisco to LA, 620km (385 miles) away, in an easy two hours, a significant improvement on the more than six hours it'd take by car in light traffic but longer than the 35 minutes Musk promised in his Alpha Hyperloop paper. However, his proposal linked county line to county line, whereas trains can more easily whip into the centre of cities. There are, of course, challenges and high costs to installing train infrastructure – just ask the UK, which has struggled to get its second high-speed line funded and approved and has already had to cut back route plans – but at least you wouldn't have to invent pods and test tunnels first, and get the whole apparatus certified by regulators.

Musk's dismantled track aside, no one else has successfully built a hyperloop track beyond 500m (1,640ft) long. And no test has topped 500km/h (310mph), let alone that promised 1,000km/h (620mph) top speed. But engineers are amazing, and the physics are sound, so one day it might well work. However, even if the technical challenges can be unpicked, practical hurdles still must be considered. Costs to buy land and install giant tubes will be high, and maintenance will be expensive. Any faults could be catastrophic, if not in terms of loss of life or injury then in transport delays (same as with railways; if there's a fault on the track, no train is going that way). Safety and security is a concern: will passengers be subjected to air-travel-style searches and queues, and how do you evacuate passengers in case of emergency? And then there's the maths: current pod designs carry up to 40 people, well below high-speed Shinkansen in Japan that squeeze in as many as 1,300 per train. You're gonna need a lot of pods.

There are positives to the idea, of course – otherwise, it wouldn't captivate so many clever people. There's two main ones with wider appeal, and a secondary one that solves a problem for governments. First, being able to travel across a country in half an hour would be incredible; imagine the life-changing joy of seeing far-flung friends as easily and as quickly as hopping on a train. And second, it would be ideal, to say the least, to be able to do so with a generally green form of transport rather than exacerbating the climate crisis by flying or driving slowly in intense traffic.*

* It's worth noting that hyperloops are only as green as their local energy mix, as they do require electricity; that electric trains do exist; and that stacking concrete or steel tubes on to concrete pylons for hundreds of kilometres will have one heck of a carbon impact.

Now, about the further point alluded to above. Hyperloop feasibility reports have a funny tendency to claim that no public funds are necessary, and that the entire project can be paid for using private funds. Naturally, that appeals to local governments – a world-first, sustainable, mass transport system, for free! – as well as to private investors hoping to rake in the big bucks. Indeed, analyst reports suggest the hyperloop market is worth well over a billion dollars already, though no hyperloop exists yet. Hyperloops are cheap when marketed to governments, but lucrative when seeking out investors – this is magical maths. But of course, all the numbers are a bit fudged: we have no real-world example to look to because hyperloops don't exist.

Someone, someday will build something like a hyperloop, if only to say they were the first. Perhaps it'll be an oil-money-rich Middle Eastern state, or China will pip everyone else to the finishing line with both maglev and hyperloop. Either way, if you want super-fast public transit now, remember that high-speed trains already exist, and pneumatic tubes, hover trains and hyperloops still don't, despite more than a century of effort.

Building better transport systems is worthy of applause. If only we invested more in improving train designs, pneumatic or otherwise, and less in developing cars. But we need to learn our lessons and be wary of adding to the pile of low-drag trains that are derelict around the world, and spend our money more wisely using systems we know work. We need green mass transit now, not decades in the future.

CHAPTER EIGHT

Smart Cities

In 2016, I was offered a last-minute press spot on a UK trade mission to Malaysia and Singapore. I was thrilled to be offered a free trip to two very expensive-to-visit places that looked (and are) remarkable and beautiful and sunny.

On the way between the two countries, we stopped off at a newly announced city, which Malaysia was building with Chinese developers Country Garden. Copying Singapore's land reclamation, Malaysia was forming a quadrant of islands out of largely nothing in the Straits of Johor, plonking 'Forest City' on top of them.* We toured the sales suite, set in a swooping white-and-glass building surrounded by lush jungle foliage, soft beaches and – inexplicably – giant crabs. Those were fake, thankfully, but so too was the two-storey green wall of plants that lined the interior of the sales building.

Forest City has never been marketed solely as a smart city. Instead, it was designed as a luxurious sustainable neighbourhood to rival Singapore and draw Chinese investors to Malaysia. But such city-from-scratch projects (or at least their marketing teams) can't help but pull in the latest future tech, whether it exists or not. Describing it at the time as 'technology driven', 'futuristic' and 'the Model of Future Cities', the sales team for the $100 billion neighbourhood claimed the buildings would be so smart that a window smashed by an errant football during the day would be automatically spotted and fixed before a resident returned home from work. The jungle of greenery cascading down every facade would be kept alive via an automated irrigation

* I wrote about this for *WIRED* at the time – you can read the story here: https://biturl.top/a226ji

system that would know exactly what each plant required. (At the time we visited, admittedly early in the city's development, there was no automated sprinkler system – hardly cutting-edge in itself, since my parents had one in their Canadian suburban lawn – but rather human gardeners standing around pointing hoses.)

The Forest City website has more dramatic, fanciful claims in a section labelled 'green and smart' – a pattern of mixing sustainability and tech used by plenty of from-scratch cities. But at its core, every single element of Forest City – person, building, product, service, whatever else, perhaps even windows and plants – would have a unique identifier to let it be tracked and analysed by an urban operating system to create 'efficient living'.

Of course, you don't have to worry about delivering prototype future technologies for a city that doesn't exist. Political challenges, financial wobbling, and the COVID-19 pandemic all intervened to dislodge the dreams around Forest City. A few years after my visit, though apartment blocks had risen from the reclaimed land, as few as 500 people reportedly lived on the Johorian islands; by 2019 that had increased to several thousand.* That's small compared to the initial plan of 700,000 residents, though given the tiny size of the reclaimed land, the latter figure would have made it the area with the highest population density on the planet, so perhaps it's a positive to have missed that particular target. One reporter, who visited in 2022, said the businesses were empty and many apartment blocks were mostly dark at night, though she did add that the golf course resort was nice and busy. Local reports suggest the wider site's ghost-town vibe is actually the main tourist draw – that and the duty-free.

* A reporter for Insider toured the project in 2022 (https://biturl.top/eQJzY3) followed by the BBC the year after (https://biturl.top/aIvyai). If anyone wants to send me on an all-expenses paid trip to check it out again, I'm game.

This isn't a criticism of Malaysia or even Forest City, but of the idea that a city can spring from nothing and add citizens afterwards. As we'll see, building a city from scratch rarely works. Just look at Songdo, Korea; Masdar City, UAE; and Google Sidewalk's plans for the Waterfront district in Toronto, Canada. Each failed in a different way – the last one never even got off the page. Urban design is hard enough without adding Silicon Valley shenanigans into the mix, and it's difficult to attract people to an empty new neighbourhood on the promise of video conferencing and sensors. There's something to be said for organic growth, and I'm not talking about vegetables.

For the most part, smart technologies are used in cities that already exist, so the challenge is retrofitting connected sensors into the existing urban landscape. The so-called Internet of Things and edge computing can help cities run more efficiently on ever tighter budgets, which is necessary as more and more people pack into urban areas, with two-thirds of the world's population expected to be living in cities by 2050. Done well, such a system can slash budgets and help keep people safe and healthy, but when done poorly such technologies raise problems, be it surveillance, flawed decision-making, or exacerbating bias and inequalities. In Barcelona, bins notify the city that they need emptying; in London, they overflow while tracking people to show digital ads. It's clear where the priorities lie.

Beyond active dangers like intrusions of privacy, it's also worth considering what can't be monitored or improved by a smart system. 'Filtering urban design and administration through algorithms and interfaces tends to bracket out those messy and disorderly concerns that simply "do not compute",' writes Shannon Mattern in her book *A City Is Not a Computer*. 'We're left with the sense that everything knowable and worth knowing about a city can fit on a screen – which simply isn't true.'

If smart cities are so inherently flawed, why are they happening? The answer lies with a cohort of companies that effectively invented the idea to get their hooks into municipal budgets, with city authorities desperate to find solutions to growing problems – and politicians who want to look progressive by riding the tech wave. Depending on the definition of a smart city, they go back a very long time indeed, though arguably what's required to be a smart city has changed over time as technology has advanced. Early smart cities, from well before the term was coined, collected and applied data to solve problems – we'll see an example from the 1850s – and though that sounds smart it doesn't always end well, especially once that data is plugged into algorithms or whacked into a dashboard with little other purpose than as a backdrop for politicians at press conferences.

★ ★ ★

It's Victorian London, and cholera is creeping through the streets of Soho, a district right in the centre of the city. Famously, Dr John Snow deduced that the disease wasn't spread via bad smells in the air, as first believed, but through water – in particular, a single water pump.

He figured that out with data and a map. The clever doctor simply plotted on a map of the district all the deaths of infected people and had the good sense to notice that they lived near the same water pump on Broad Street. As the story goes, he ended the outbreak by removing the pump's handle. Dr Snow also spotted that no workers from a local brewery died of cholera, likely because they drank beer rather than water.*

That's the data collection and analytics side, one aspect of smart cities. The other aspect is controlling people (and

* There is a pump commemorating the event on the corner of what is now Broadwick Street and Lexington Street in London's Soho — as well as a John Snow pub.

whatever else) through technology. For example, the world's first traffic light was installed in London just three decades after John Snow's data map beheaded that Broad Street pump, at Bridge Street near Parliament Square in 1868. This was before cars, but a thousand pedestrians were still being killed each year on the city's roads thanks to carriages. The signal wasn't particularly smart: the 6m (20ft)-high light was manually operated, with gas-powered lights. The traffic light lasted less than a year, when it was taken out by a sub-pavement gas pipe explosion that injured the policeman operating it.

Despite such explosive origins, the idea eventually caught on and spread around the world. The first such traffic control arrived in Paris in 1912, operated not by a policeman but a police*woman* – progress! The same year, Salt Lake City, Utah boasted the first electric traffic light, followed five years later by an interconnected system covering six intersections that could be controlled simultaneously with a switch. Automation was introduced in LA, California in 1920 with signals that used timers, an idea followed by Wolverhampton, UK in 1927, while the three-light signal was invented by African-American Garret Morgan in 1923.

Automated traffic lights were seen as a technology that cut costs, particularly on labour, but they also introduced the idea of machines to control what people do in a city. By 1928, the use of automated signals allowed New York to slash its traffic police from 6,000 officers to 500, saving more than $12 million in the process, though of course electricians needed to be hired. But it also meant pedestrians and drivers had to obey the directions of machines. This was welcomed by some as traffic cops of the day displayed bias in how they manually operated controls; automation lacked the racism of the cops who wouldn't let certain groups of people cross safely.

The rollout of automatic traffic lights continued globally, with London and Tokyo following suit by 1931. Automation required that basic road data and traffic technology be combined, which allowed for the development of so-called 'green waves', or holding all the signals going in one direction for a set of cars so they can just roll on through without stopping – the introduction of this staggered system doubled commute speeds along Washington, DC's Sixteenth Street in 1926.

Such traffic systems were largely deployed the same way regardless of the country or city. This wasn't for regulatory reasons or anything to do with the technology itself, but because of the sort of people setting up the systems. Hire a professional traffic engineer to fix your congestion, and they do the same thing over and over, often with the aim of applying ideas like scientific management to bring order to chaos.

And before such consistency, there was chaos: in New York, Fifth Avenue used orange to indicate 'go', causing confusion, as green was the default on all other roads; Philadelphia tried to use a single white light to manage one busy road and all its crossings; and Paris used only red lights. The order of the lights was also disputed, along with their meanings: China wanted red to mean go, while in an Irish-dominated neighbourhood of New York, green had to be on top of red because otherwise it was deemed to symbolise the British dominance of Ireland.

The idea that traffic management was scientific was furthered by the use of studies including surveys to gather evidence that the methods worked. But, as Clay McShane notes in his history of traffic controls, too many studies claimed that traffic improved by exactly 50 per cent, and such precision suggested rounding up and guesswork rather than accurate counting. If you're going to fudge the numbers, go with something less obvious, at least.

As cars took up more and more of city roads, authorities knew they needed to do more to address traffic. By 1915, the first traffic consultancies were set up, and 10 years later Burton Marsh was hired as the first municipal traffic engineer, working for the city of Pittsburgh. By the 1950s computers had arrived, which transformed the management of traffic and cities. The first computerised road traffic control system arrived in Toronto in 1963, set up by Josef Kates' consultancy KCS[*] and its Traffic Research division, run by Leonardo Casciato. 'Toronto has a traffic cop with arms seven miles long and getting longer,' explained one report from the local *Star-Phoenix*. 'It's a computer that already controls more than 500 traffic lights in an 80-square-mile area.'

Wires connected to sensors were buried in the road ahead of intersections, tracking the speed and direction of each car that passed using electrical induction, and sending that data to the central computer for processing to decide how long to keep the light green. Here's how it worked at one intersection, the famous and busy Yonge and Bloor. Normally, the intersection cycled through the lights in a 70-second cycle, 35 seconds on each. But when traffic was heavier along Yonge, the computer would see this, consider different factors such as weather and number of lanes, and tell the light to hold green for 45 seconds instead.

The $4 million system handled 1.25 million daily trips, with the city aiming to speed up rush-hour traffic by 15 to 20 per cent, according to local media reports. It worked even better than that: trials showed a 38 per cent boost in speeds in downtown Toronto. Accidents were apparently down 7 per cent on computer-controlled roads, but up 6 per cent elsewhere, which traffic commissioner Sam Cass attributed to 'psychological' factors: 'On streets where the system is not in use, motorists get impatient and their actions lead to more accidents.' Better to roll it out everywhere, then.

---

[*] Josef Kates is a Canadian consulting and tech success story. Born Josef Katz in Vienna, 1921, his family emigrated to Italy and then England after Germany annexed Austria. The British held him as an 'enemy alien', eventually deporting him to Canada, where he was also interned as such. Eventually they realised the mistake and he was allowed to finish his education. Katz went on to create the first digital game, *Bertie the Brain*, which used vacuum tubes of his own invention. He later became chairman of the Science Council of Canada and was working on transport improvements into his nineties, before passing in 2018 at 97.

But what happens when such a system fails? Toronto found out when the traffic lights started operating at random, and engineers couldn't unpick why. Eventually, they figured out that fluctuating voltage meant incorrect data was being fed back; the system understood the data was wrong and rather than give the traffic signals bad directions, it shut itself down. And as the system aged, it failed more frequently. Rather than replace it outright, in 1980 an additional system was installed using PerkinElmer computers and chip-based sensors instead of the more basic electromechanical wire coils. The new system also added a human into the loop, who would watch over the system via a 6m-by-3m (20ft-by-10ft) map of Toronto covered in LEDs. It's not a smart city until you have a dashboard with blinking lights, after all.

Traffic is an easy system to computerise – count the cars, analyse their speed and leave lights green a bit longer. But, of course, it didn't take long before city authorities decided they could manage much more than traffic jams with computers. This wasn't just about computing power, but how the idea of computing changed how we thought about the world: we started to look at cities as flows of information and people as problems to be solved with models and algorithms. Data was knowledge, computing was science, and whatever the models spat out must then be correct.

In 1969, Jay Forrester's *Urban Dynamics* laid out a scientific way of considering cities as systems based on industrial dynamics, which considers the feedback loops and interactions in a network to understand how to make it work better. It sounds clever, but cities are more complex than such models allow for and in the end he treated all cities the same when they are very much different places. If you wanted to solve unemployment, for instance, his models didn't suggest training or government support, but clearing out slums in place of more expensive housing – in other words, notes one researcher, gentrification. Trying to systemise cities in order to collect relevant data that can be analysed to power automation is a flawed way of thinking: there's plenty about a city that can't be collected in the numerical format necessary for a computer to analyse. Not everything can be automated – we've had traffic lights directing cars for more than a century, but lollipop persons still help children cross safely outside schools. Analysing how a city works and what solutions it needs also can't be done from one angle. This is why we have elected mayors and councils full of people; deciding what to prioritise is deeply political.

LA gave such data gathering a go in the 1960s, notes academic and urban planner Mark Vallianatos in an article for Gizmodo, as part of a wider plan to build an Urban Information System to address everything from poverty to land use*. In 1967, the city officially formed its Community Analysis Bureau, aiming to give decision-makers insights to help prioritise LA's needs.

The city was broken up into 727 census tracts (neighbourhoods or districts) with clusters based on 66 pieces of data to describe them, including ethnicity, crime rate, income, housing and more. For example, 'post-war suburbs' was a cluster that applied across 76 census tracts; it's largely middle class, middle-aged, and white, with income and education levels just above average; they tend to own cars and not take public transport. Another cluster is 'singles': younger (average age 33) people living in higher-density apartments, with a high level of education and an even higher level of labour participation (i.e. jobs) – it's your '*Sex and the City*' crew. And then there's 'high income' – mostly white people, with large homes, and incomes three times as high as any other cluster. But there were surprises: this cluster, which was found in Encino and Brentwood, also had the highest concentration of unrelated individuals in poverty, double the city average. The report pins this on low-salaried 'live-in servants', or a statistical error.

Suburbanites, single professionals, rich people: those are all fairly obvious. Other clusters included 'The Chinese', which

* You can read Vallianatos' article here: https://biturl.top/q2Ib6v

unsurprisingly includes Chinatown; the 'Once Elegant', a mix of elderly people and younger Hispanic families; and 'Blacks of Moderate Income', which is 75 per cent populated by Black people, has a high rate of female workers, and low crime. 'This is the first time this technique has ever been used to describe a city,' wrote Robert Joyce, the CAB director, in its first report.

Data sources informing those clusters were varied. The team digitised Census Bureau data, but also used aerial photos to assess the physical environment of a district, with staff scoring it based on characteristics such as litter, vegetation and whether homes had 'convenience structures' such as patios and swimming pools, notes Vallianatos. Those scores were checked by driving through a neighbourhood to look for signs of neglect, unquestionably a problematic way to collect data that introduces bias and prejudges communities without any real insight or investigation; graffiti and untrimmed lawns are signs of neglect in some neighbourhoods, and evidence of hipster artists and no-mow-May in others.

The goal of all this was to figure out which neighbourhoods needed help, and what sort of help. Vallianatos notes that the bureau didn't actually need all 66 data points that were used to define the clusters. Just three acted as effective red flags: age of housing, average birth weight of babies and sixth-grade reading scores. But here's where it all so often falls down: now it was up to politicians to take action. Handily, one thing that data analytics are good at is winning grants – how do you say no to scientific evidence, after all? Hyper-local grants were doled out to improve streets and parks, expand social services and so on. In the end, the CAB became a way to get grants, Vallianatos notes, rather than a way to inform decision-making. As computers became more common, the CAB was eventually folded into wider planning divisions, shedding the original motivation to add to the 'quality of life; for each and every citizen of Los Angeles'.

* * *

Over on the other side of the country, New York City was burning down. This is no exaggeration: between 1970 and 1980, census data shows 40 per cent of the buildings in the South Bronx were lost to either fires or abandonment. Some areas lost 97 per cent of their buildings. The Bronx was clearing out for a multitude of complex reasons, but it's no surprise people don't want to live on a street of burned-out shells, with some streets entirely levelled as though bombs had been dropped.

During the second game of the 1977 World Series of baseball, an abandoned school went up in flames visible from Yankee Stadium; no fire truck showed up, giving the audience a front-row seat to the reality of life in the less wealthy parts of New York. Fires raged across Brooklyn, too: the night of the notorious 1977 blackout, more than 1,000 fires were reported across Bushwick; and on the night of 6 March 1970, seven children – five from one family – died in a fire in Bedford-Stuyvesant. According to Joe Flood's book *The Fires*, while much of the blame was pinned on arson, that was only ever a small part of the problem.* By the mid-1970s, the city was out of money and more than $12.5 billion in debt. Years of neglect meant building regulations weren't followed and there was a widely held belief that authorities were happy to see entire neighbourhoods burned to the ground in the hopes they could rebuild something better. Fire stations were poorly staffed, leading to closures, and many fire companies the city chose to shutter were the very ones at the centre of communities that were burning down.

* As Flood notes, during the 1950s, less than 1 per cent of fires were caused by arson. That number didn't rise above 1.1 per cent until 1975. At its peak in the late 1970s, arson made up less than 7 per cent of fires. 'What's more, arson occurred primarily in already burned-out, abandoned buildings – after all, it made more sense to torch a building without rent-paying tenants than one that had at least some revenue coming in.'

How did this happen? At the time, John O'Hagan had just taken over leadership of the New York Fire Department (FDNY). Flood reports that O'Hagan was systematic and methodical: his fire safety efforts and innovations slashed fire fatalities nationwide by 40 per cent.* Under his watch, the FDNY was the first to use computers to track fires to spot trends, and as a man who believed in methodology and systems, O'Hagan listened to RAND – and that proved to be a well-intentioned mistake. Consultants at RAND made a model to decide which fire stations to close, and that model was wrong, causing fire deaths to skyrocket and a failure to maintain key infrastructure such as hydrants.

RAND Corporation was founded in 1948 by the US Air Force and Douglas Aircraft Company, a defence contractor. The aim was to maintain hold of research and development after demobilisation following the Second World War, given that all the scientists and experts the war effort had drawn together were going to disperse.† By the 1970s, RAND was looking to bring systems analysis approaches to cities, 'the kind of streamlined modern management' used by the US Secretary of Defense in the Vietnam War (because that went so well for the US).

New York teamed up with RAND on its own bespoke think tank, pulling in consultants to run the numbers and build a model to inform how the city should be run. When it came to fire companies, RAND looked at response time data to figure out where to add companies and where to remove

* He didn't necessarily invent everything, but took on smart ideas like a steel bar called a Halligan that made it easy to get into buildings, eye shields for helmets, a U-pipe for getting water into high-rises, and a jaws-of-life-style saw. He also helped develop an infrared system for spotting fire in walls.

† A few facts about RAND: yes, it does stand for 'R and D' – as in, research and development; additionally, it was the model for the BLAND Corporation in *Dr Strangelove*, and is widely considered to be the first think tank.

them. However, the model had outright mistakes, Flood shows. It ignored traffic's impact on response time and assumed fire trucks were always available – in busy areas, they were not. That meant an in-need area of the Bronx might look overserved compared to brutally underserved neighbours. As Flood notes: 'There's an old modeller's dictum that's the rough equivalent of "missing the forest for the trees": solving the equation doesn't necessarily solve the problem.' In short, RAND recommended shutting fire companies in the very neighbourhoods that were burning to the ground.

Largely the model's output largely reflected O'Hagan's existing biases, meaning he felt empowered by the algorithmic evidence to slash at will. Amid rising deaths and out-of-control fires, RAND's model continued to recommend cuts to the worst-hit neighbourhoods and busiest stations. No one needed a chart or statistics to see that was the wrong decision, but when faced with men in suits pointing at figures, we tend to lose our common sense.

To be clear, none of this was done with any apparent malintent. Cuts were needed. But this early version of data-based systems analysis utterly failed.

In 1972, when the cuts came into force, they were immediately criticised. O'Hagan's predecessor joined a lawsuit claiming that poor communities of Black and Hispanic people were hardest hit. Newspaper journalists noted that the closure of these fire companies didn't even save all that much money, while one pointed out that an area that had lost a fire station had seen 82 fire deaths in the preceding seven years. The reality clearly didn't reflect the RAND data, but because the fire authorities were using an analytical model, the judge in the lawsuit said the closures couldn't possibly have been based on race: 'The specific decision as to which fire companies would be eliminated, was premised solely upon the neutral, nonracial, scientific and empirical data available,' wrote US District Court Judge John Canella in his decision. (We now know perfectly well that algorithms do reflect racism and other biases; in this case, the vast majority of stations closed were located within poorer, ethnic minority neighbourhoods, suggesting the model prioritised the safety of people in other areas, though not intentionally.)

In 1975, New York City's budget cuts hit RAND itself, and city authorities ended the contract. Its legacy was clear: the number of serious fires went up by 40 per cent between 1974 and 1977. Some stations ran out of fires to fight, as there was so little left to burn in their districts. By 1977, O'Hagan was out as commissioner and chief after his political allies lost control of the city in an election. Numbers can lie, ruin a career and even burn a city down.

* * *

All that is really the pre-history of smart cities. As a formalised idea, using tech and analytics to run cities really came into its own because of Bill and Hillary Clinton; well, their foundation anyway. In 2005, the Clinton Foundation handed networking giant Cisco a challenge via its Clinton Global Initiative: figure out how to make cities sustainable. (The connection between 'smart' and 'sustainable' began right from the beginning, as the two are frequently conflated, despite referring to wildly different ideas, likely a result of marketing teams.)

That sparked the Connected Urban Development programme – you can see why the industry ended up with the name 'smart cities' in the long run, as CUD is a terrible acronym – as well as a series of conferences. Three cities – Amsterdam, San Francisco and Seoul – were chosen for pilot programmes to develop technologies to slash carbon emissions, with the aim of creating techniques applicable in other cities. What sort of clever ideas did they come up with? In Amsterdam, they invented co-working spaces centered on Cisco's TelePrescence video conferencing system, saving on carbon emissions from commuting. In Seoul, Cisco built a web app for planning travel that showed carbon emissions for each route. For San Francisco, it built a digital map to track progress on carbon reduction goals, as well as setting up a Living Innovation Laboratory at the Hunters Point Shipyard residential redevelopment – though health concerns over the former nuclear test lab on the site and fraud delayed that project. Building homes on a nuclear dumping ground is neither smart nor sustainable, after all. That smart city lab doesn't appear to have been built.

Now, it's no surprise that if you ask a networking company to fix cities, it's going to recommend solutions that involve a lot of networking equipment. Ask Cisco to solve urban woes, and it'll recommend you install superfast broadband to access its video conferencing system. The travel app does sound properly ahead of its time, at least. After the five-year partnership with the Clinton Foundation ran its course, CUD was handed to The Climate Group, an environmental non-profit, and Cisco evolved the benevolent work into commercial products under the Smart and Connected Communities name, though later also began using the phrase 'intelligent urbanisation'.

IBM had been pitching the idea of urban tech previously but started using the term 'smart city' in 2009. Before that, there was barely any mention of the phrase; after that point, its use skyrocketed. That year, researchers Susanne Dirks and Mary Keeling discussed the necessity of smarter cities in a white paper for the IBM Institute for Business Value, writing: 'to seize opportunities and build sustainable prosperity, cities need to become "smarter".'

Of course, projects based around similar principles, such as the Korean city of Songdo – which we'll find out more about in a bit – existed before this, but the idea of a 'smart city' goes beyond just building a modern city from scratch, or rethinking urban planning, or using technology to better organise or manage services. Instead, 'smart city' is a marketing tool – and 2009 marks a tipping point, where the idea shifted from being led by municipal governments to becoming a service brought in by consultants. With that IBM paper, tech suppliers took the idea of governing by consultancy and absolutely ran flat out.

To be fair to Dirks and Keeling, their argument is sound: technology can offer better operational controls at lower costs. And it was increasingly necessary, as in 2008, for the first time, most of the world's people lived in cities versus rural areas. That added stress to transport systems, hospitals, trade, and utilities such as power and energy. According to Dirks and Keeling, that required cities to become 'instrumented, interconnected and intelligent' – that simply means digitising, connecting it all up via the web, and analysing everything – while recognising that cities were interrelated 'systems of systems', harkening back to Forrester's work on urban dynamics and systems for cities, which we explored earlier.

The first major IBM 'smarter cities' project took place in Rio de Janeiro, Brazil, beginning in 2010 as part of the preparations for the country holding two massive sports events, the 2014 FIFA World Cup and the 2016 Olympic Games. The centrepiece of the project – or at least the thing that got mentioned a lot in media coverage – was the control room. According to the *New York Times*, it looked straight out of NASA, while *the Guardian* compared it to 'a Bond villain's techno-lair'. The operations centre had a massive wall four monitors high and 20 wide, showing weather predictions and mapping car accidents and power failures, as well as displaying video streams of train stations and key intersections. Around 70 employees watched this dashboard from banks of desks with their own displays, all wearing white jumpsuits – apparently to encourage solidarity rather than just because it looks cool.

Of course, watching a video feed of a busy station doesn't help you do much about the crowds, so IBM pulled together relevant data, and algorithms looked for patterns. Guru Banavar, IBM's CTO for the public sector at the time, told the *New York Times* it was 'sense-making software'. Rio needed that, then-mayor Eduardo Paes said, after a landslide and flood had hit the city and he realised there was no way to see an overview of what was happening, making it difficult to decide how to respond. Paes needed to show to the Olympic organising committee and FIFA that his city was under control in emergencies; indeed, a secondary control centre next door in Rio and another in Brasilia were contractually required under the FIFA deal.

While it's neat to see where all your buses are and handy to manage emergency response in case of a disaster, a real-time dashboard only goes so far when it comes to addressing the daily problems facing most people. A study of the Rio smart city system by University of Zurich researcher Christopher Gaffney found it 'mostly lacking' – in large part because the whizzy dashboard didn't address 'problems of radical inequality, or systemic poor governance, or compromised urban planning agendas – all of which continue to be the "dumbest" elements of Rio de Janeiro'. The system had limited coverage and functions, despite the big displays.

Plus, the analysis noted that the out-of-the-box software provided by IBM wasn't always suitable, and collaboration was difficult, meaning Rio city employees largely wrote their own software. 'As a result, there is no reference to IBM in the [operating centre], yet as part of an agreement between the two parties, IBM can use Rio de Janeiro in its Smarter Cities advertising campaign,' Gaffney's paper notes.

Much of the smart city dream is illustrated and epitomised by dashboards.* They are modern and futuristic, offering a perfect backdrop for media interviews during disasters to show that everything is under control, as Rio's mayor liked to do – it is, after all, a *control* room. On the other hand, city staff only consider what is on the displays, what can be counted

* Shannon Mattern notes in her book *A City Is Not a Computer* that the word 'dashboard' was first used in 1846 in reference to a board at the front of carriages that kept mud from being kicked into the cab by horses. It didn't begin to mean a display screen until 1990, though did refer to speedometers and other car dials before then.

numerically and plugged into an algorithm, literally leaving more complex aspects of city life out of the room where decisions are made.

Rio wasn't the first city to feature data displays for politicians to study. Salvador Allende had a hexagonal 'Opsroom' in Santiago, Chile in the early 1970s, built as part of the Cybersyn project,* which pulled in economic data via telex to displays that an operator could switch between using buttons built into futuristic-looking chairs.

While Rio's room-sized dashboard beguiled at first, it soon became clear that even when such a setup technically works, it's often little more than just displays of charts and video feeds, rather than actual smart analytics of a city system. But they often *don't* work. Then-deputy Antoni Vives helped build Barcelona's smart city system and toured other cities to see their projects, visiting an unnamed South American city's situation room with the local mayor. 'It was obvious that it hadn't been used in months: dust on the tables, disconnected screens, ageing computers,' Vives wrote in his memoir of the Barcelona project. 'The danger of large companies on the hunt for an unsuspecting mayor is that they might fill the streets with sensors that don't do anything, and offices with screens that do even less.'

Or, they might build cities out of nothing that no one wants to live in.

⋆ ⋆ ⋆

Along the edge of Incheon, South Korea, a new district on a 6-square km (2.4-square miles) island was being planned.

---

* Adam Greenfield in *Against the Smart City* notes one thing the Cybersyn project did get right: it not only included the rights of citizens to be stakeholders in decision-making, but acknowledged that the control system belonged to *el pueblo* – the people – and that they should have access to it. That's not a provision common in most smart city designs.

Announced in 2004, Songdo New City was unveiled as a Korean rival to Shanghai, Hong Kong or Singapore, beginning with a conference centre. By the time of its projected completion, around 2012, the project's costs had risen to $35 billion, with canals mimicking those of Venice, a park modelled on Central Park in New York City, and an opera house that nods to the one in Sydney, Australia. It aimed to house 300,000 residents working in sectors like IT and marketing across a business centre the size of downtown Boston.

A few years before IBM started using the 'smart city' term, Korea had its own: 'ubiquitous city' or U-city. And that idea would be applied to Songdo by John Kim, a former Yahoo designer who had been approached to build tech into the new city. Songdo would be a testbed for new tech but also show off what life could be like when everything is digitised, or, as they called it, U-Life. That included smart rubbish bins that sucked garbage into tubes to be recycled, streetside computers for anyone to use, and smart homes with phone-controlled lights and heating, as well as intelligent road pricing (automated toll roads, basically), smart ads that changed to reflect who was looking at them, and automated parcel delivery.

In 2005, while buildings were still being planned in Songdo, Kim promised a world organised around a cutting-edge new technology: RFID cards. As we saw in the chapter on cyborg technology, radio frequency identification is the wizardry that lets you wave a card with an embedded chip to pay without typing in a PIN. That would be used by residents not only to open their front doors, but also to access transport such as buses or bike shares, pay parking machines or even just see a movie, Kim said at the time. He told the *New York Times* that the system would be anonymous, and cards could easily be cancelled if lost to gain access to your own front door.

Songdo also promised citizens would have the ability to make video calls, stream video content anytime and have access to their digital content anywhere in the city. Kim's description of Songdo is the life many of us lead now, we just didn't need to dredge the sea to build an island first. Sure, it's easy to laugh at these plans as naive given what happened as mobile connections improved, but for context, at this point Netflix was still mailing DVDs rather than streaming movies on demand (it didn't start doing that until 2007).

In 2005, the plan for Songdo was to build the technology infrastructure locally and look to overseas tech companies for support as partners. In 2011, Cisco signed on, beginning with an investment in a new joint venture called Songdo U-Life, including the use of Cisco networking to connect the island's buildings. The press release promised more than 10,000 Cisco TelePresence units for video conferencing, so residents could access education and concierge services, as well as sensors embedded in buildings to track energy use. This would allow residents to live a 'technology-enabled lifestyle through digital infrastructure systems designed and built directly into the city's framework'. That said, the 'smart' technologies of Songdo only made up 2.9 per cent of its budget – it's a small charge to make your city seem futuristic.

So what happened? After 15 years of work, it was clear Songdo wasn't the draw for residents nor investors that its creators had hoped, with only a quarter to a third of the expected 300,000 residents moving in. Journalists described it as a 'Chernobyl-like ghost town', with residents saying it was like 'living in a deserted prison', while a local blogger said, 'future South Korean zombie movies could film here without having to worry about anyone getting in the way'. Others say it depends on where in the city you go, and you can find outposts of liveliness among the otherwise quiet streets; it's perhaps more, some say, like an American suburb than the busy heart of Seoul.

The problem appears to be twofold. First, not enough companies rushed in to set up business, giving people little reason to live in an empty area that's more expensive than neighbouring ones. And second, streaming movies, video calls and smart homes are all widely available – there's no need to move to a lonely 'smart city' to access such luxuries.

This is the inherent problem with building a smart city: the 'smarts' are often out of date before anyone moves in. The streetside computers, video conferencing and RFID tags that sounded cutting-edge when Songdo was first planned are useless now that we have smartphones in our pockets.

Digs about zombies and out-of-date tech aside, it seems the mistake was building Songdo all at once, rather than letting it evolve organically, while locking in technology that quickly aged. The failure to draw companies and residents gives Korea a chance to take a breath and rethink what to do next – and slow and organic is a better way to grow than fast and top-down. The plan has been rejigged to attempt to build the city into a 'bio hub' to attract more biotech companies, as well as turning it into a testbed for smart city tech. The problem with testing in a ghost town is it's more of a lab than a city; Songdo may well be a good place to trial a smart bin, but you're going to have to try it somewhere with people next.

* * *

Around the same time Songdo was emerging from the sea, the UAE unveiled in 2006 its own city in the desert: Masdar City. Set on the outskirts of Abu Dhabi, the 6-square km (2.4-square miles) zone was designed in part by Foster + Partners to house 50,000 residents but also welcome 40,000 commuters daily. Described at its launch as the first zero-carbon city, it would show off sustainable technologies from desalination plants to solar panels and even a 45m (100ft)-tall tower to pull cool winds down into the inevitably hot streets, with walls around the city to block the desert winds and noise from the airport next door. Masdar City was a smart city with the emphasis on sustainability.

As part of the city's role as an urban lab for clean tech, an autonomous personal pod system would ferry people around to prevent car use, and a computerised navigation system built into pavements would help walkers find their way. Construction began in 2008, and Foster + Partners picked up several sustainability awards for the design. 'We want to position ourselves as thinkers and progressives,' Sultan Al Jaber, CEO of the company behind Masdar, told reporters at the time. 'Years ago in the Middle East, we lived in a very sustainable environment. We are bringing that back by creating a compact city where people don't need to use a car.' It all sounds fabulous. Plans often do.

Originally scheduled to be completed within 10 years, by 2016 Masdar City was only 5 per cent built, with a new deadline set for 2030. The handful of buildings are nowhere near the carbon-free goal, which has since been rolled back, while the autonomous public transport system has been scrapped. It's now being described as 'the world's first green ghost town', with just 300 people living on site, all students who were given free housing. The Foster + Partners master plan which won so many awards was ditched in 2010, with the building work continuing in a more ad hoc manner.

French photographer Etienne Malapert captured images of the desert neighbourhood, which showed rows of parked cars, belying the truth of the environmental transport plans, and workers toiling to keep foliage green.* The Google satellite view of Masdar City, using imagery from 2023, shows more space set aside for solar farms than buildings,† greenery-lined roads running along empty desert, multiple large parking lots – but also a massive park with a playground that looks epic from space.

What went wrong? Building a city is hard, even if it is really just a suburb between a city and its airport, as Masdar

---

* You can view Etienne Malapert's photography here: https://biturl.top/r6Vv2y

† The solar panels were apparently initially meant to go on the rooftops, but dust interfered with them. Placing them all together on the ground makes them easier to clean, which is a sensible solution and highlights the difference between planning a city and building one.

City is. And building it to be car-free in a car-loving country, carbon-free in an oil nation, cool and walkable in a desert, suggests the designers hadn't really thought things through. As anthropologist Gökçe Günel described in her book *Spaceship in the Desert*, Masdar City was a 'spaceship insulated from the rest of the world', based on a vision 'that the desert is an empty zone on which any kind of ideal can be projected'. And that simply isn't true.

No one called this a smart city, but alongside the autonomous pods, Masdar City promised 'an image of the future drawn from science fiction,' Günel writes, noting those working on the plans said it would be 'smart' with 'a hidden brain' that would know when you return home in order to cool your flat before you arrive, with data about the environmental performance of the city complex shared on large screens in public places as 'uplifting news' for residents.

The autonomous pods were perhaps the most sci-fi, future-tech aspect of the Masdar City plans, so it's no surprise that they're the aspect that was first to fall by the wayside. The Personal Rapid Transit (PRT) system was described by *Time* magazine as looking like something out of the film *Tron*. Driverless and electric, these cars were to find their way using tiny magnets hidden in the road – this will sound familiar from the driverless cars chapter; it's an idea that finds new fans every decade – supposedly negating the need for traditional cars within Masdar. However, as one journalist reported in 2011, there was just one station, with a single stop 800m (half a mile) away: 'I could have used a much older form of transport: my legs.'

That barely-a-network falls rather short when compared to the plans: 87 stations with 1,800 passenger pods, as well as a secondary freight system. The aim was for the magnet-controlled pod network to run beneath the city in a layer dubbed the 'undercroft', so it wouldn't disrupt pedestrians. To achieve this, Foster + Partners' plan called for the city to be built 6m (20ft) above ground level. Unsurprisingly, this was deemed too expensive and the PRT system was ditched in 2010. 'The actual pod car and track technology had to be invented from scratch, and it had been difficult to predict costs and to streamline the production,' Günel writes. Just like hyperloops, eVTOLs and other new forms of transport, if you're building now, you need mass transit systems that exist now.

So, the PRT was cancelled, but not before the single-stop route between the Masdar Institute and its car park was built, offering a free 2½-minute journey. Buses now operate throughout the parts of the rest of Masdar City that are built, and car parks abound – it's hardly the vision of the future that Masdar City and Foster + Partners sold.

But perhaps the PRT served its purpose. In the early days of the project, it was the PRT that was often trotted out to highlight the progressive, innovative nature of Masdar City, writes Günel: 'The pod cars confirmed that humanity's future would be one of technological complexity, just as in science fiction movies.' And, just as we've seen the embedded-magnets system in roads before, the idea of PRT isn't new: that's effectively what motorway driverless cars are, letting us join up with 'trains' of cars at higher speeds without having to interact with other people or switch to another form of transport for the last mile home.

Similar ideas were suggested in Paris in 1967, while a 'Cabinentaxi' system was built in Hamburg, Germany and a computer-controlled vehicle system set up for an expo in Osaka, Japan. Safety concerns scuppered the Japanese project after six months, while Hamburg's trial lasted until the 1980s but then went no further due to budget concerns. The Paris project never passed regulatory approval. There is one place you can see such a system in operation, however: at London Heathrow airport, where driverless pods run on a track to ferry people to Terminal 5 from the car park – not wholly different from Masdar City's eventual small-scale network at the Masdar Institute.

Günel described using Masdar's experience of the PRT pods: the slightly uneven concrete in the roads caused problems for the suspension, increasing the maintenance costs; because the route was so short, the top speed of 40km/h (25mph) was unnecessary and impossible; and some students preferred the shuttle bus because when the pods break down, they rely on someone else to come and open the doors, leaving them waiting 20 minutes in the heat. An expensive toy, she quotes one student as saying.

Masdar City hasn't met its carbon goals, has failed to fully develop automated personalised transport and remains largely unbuilt – Abu Dhabi's leadership, which owns all of the land and buildings, is apparently only constructing new properties when a company steps up to lease them, which is actually a sensible way to operate. The plans that were shelved due to costs, wider economic challenges and technology issues have instead turned to a slower evolution that has yielded some benefits. Geothermal wells have been installed to test their viability as a sustainable energy source and the buildings that have been constructed are unquestionably better suited for hot desert climates than the glass-covered skyscrapers of neighbouring Abu Dhabi, though reports suggest Emiratis aren't yet keen on moving to the futuristic city.

This isn't the future Masdar City promised. Indeed, the head planner at Foster + Partners was unable to tell one researcher what the neighbourhood would look like if ever finished. We'll all have to wait until 2030, and likely far beyond, to see what is eventually built.

⋆ ⋆ ⋆

Sidewalk Labs was quietly formed in 2015 to give Google co-founder Larry Page's 'urban innovation' ideas – no one wants to call them 'smart cities' any more – a place to live under the umbrella company, Alphabet. Core to his thinking was a desire to rebuild cities without having to deal with regulators and city authorities; Page even joked that building on the moon would be easier than negotiating with municipal governments, according to Josh O'Kane's *Sideways*, a detailed look at the short-lived company's Canadian project.

Indeed, pre-Sidewalk, Page's urban solutions included building a libertarian city-state on a ship with infamous and controversial tech founder Peter Thiel; installing a monorail with individual pods to better connect the University of Michigan; and installing a massive dome over a city for guaranteed weather, air quality and so on.* Bonkers schemes can spark smart solutions, so there's nothing wrong with sketching out silly notions, especially when you've got hundreds of millions of dollars to play with as a co-founder of Google. But the problem is finding a city willing to experiment on its own citizens – and with its own budget.

After nabbing former New York deputy mayor Dan Doctoroff as CEO, Sidewalk's first project was to task its employees with building a document they dubbed the Yellow Book. This was essentially a heavily researched selection of urban solutions and futuristic visions to present to Page, which included bathroom mirrors that analyse residents' faces for illness, a Yelp-style review system for police officers, and rewards for good behaviour based on detailed data collection, all under a bubble in a Google-built town. After a year of work, the plan was presented to Page, who seemed blasé about it, with the tech exec instead suggesting his own ideas, including buildings on wheels. Buildings. On. Wheels.†

---

* Two of these ideas have featured as jokes on animated show *The Simpsons*. In Season 4, Episode 12, 'Marge vs the Monorail', a con man sells their town of Springfield an unnecessary train. And a dome covers the town in *The Simpsons Movie*, to protect others from polluted Springfield rather than the other way around.

† Truly nothing is new, not even wild ideas like this. In the 1970s, a group of radical architects known as Superstudio created a photo series dubbed Continuous Monument, tracking a moveable building as it goes where it'd like. Yet this was a satirical warning against thoughtless expansion of cities, not an idea to borrow. More here: https://biturl.top/7rUBfi

Sidewalk then started searching for a city, or part of one, that it could rebuild and gentrify. The edges of Denver, Colorado and an old naval base in California were both considered, as were the emptying streets of central Detroit, Michigan. O'Kane reports one consultant, Anthony Townsend, recalled a swathe of the fading Motor City being deemed suitable, but upon his own closer inspection, realised it would have required the bulldozing of a dozen historic churches. That's a tough sell even for Google.

That left Sidewalk as a city-fixing startup with no sandbox to play in – until Doctoroff got a phone call from a former colleague asking if the company would be interested in a project in Toronto. The Canadian capital had a small plot of land, just 0.05-square km (0.02-square miles), known as Quayside, a little slice of a wider regeneration plan tearing up ports and industrial warehouses along Lake Ontario in favour of urban residential, parks and more. Toronto is short on land and short on housing, so Waterfront Toronto, as the project was known, was ideal for a smart, new, higher-density development from Sidewalk.

This could have been a perfect pairing: the city gets investment from a wealthy company and gains a forward-thinking neighbourhood with plenty of tech toys to draw residents and businesses, while the Silicon Valley giant has a real-world lab to trial its ideas to sell on to cities around the world. Instead, it was a two-and-a-half-year-long waste of everyone's time. Nothing got built. No tech was developed or tested. Some ideas were sketched out, but nothing beyond planning proposals. Songdo and Masdar are failures as smart cities, but they do actually exist: there are roads and buildings and a one-stop automated pod network, even if that's a far cry from what was promised. Sidewalk's vision for Quayside never left the page, held back by privacy concerns, ownership debates and land feuds, but most of all by how difficult the tech giant found following regulations, procedures and bureaucracy. (In the end, Alphabet and Sidewalk blamed the changing financial situation owing to the impact of the COVID-19 pandemic for ending the project, but it was looking doomed well before then.)

The backlash against Sidewalk taking on Quayside began almost immediately, not helped by a rushed-through announcement that raised the ire of the project's own board members, which was exacerbated by Prime Minister Justin Trudeau claiming that the deal had been in the works for years, undermining Waterfront Toronto's own – accurate, to be clear – claim that there had been an actual competition for the project.*

Before we get into why so many different people hated this project for so many different reasons, first let's see what Sidewalk actually wanted to do with this tiny slice of Toronto waterfront. Initially, Sidewalk dithered on releasing the full proposal and had to be forced into doing so. The 1,500-page document† included ideas such as covering the district with sensors to track everything from people to air quality, building everything out of wood (this would involve a new sawmill elsewhere in the province), underground tunnels for autonomous mail delivery carts (a delightful mashup of Masdar's PRT and London's 100-year-old LPDR), driverless cars (whenever they're developed) and heated paving slabs. Other ideas included robot trash collectors, automated meters to track waste and incentivise recycling, and dynamic roads that would respond to traffic patterns, switching a bike lane

* That sparked journalist Josh O'Kane into a years-long public information data request battle to find out about that years-ago call to Schmidt, which was inexplicably expunged or left out from records, contrary to the government's own lobbying rules. There was no reason to hide this information, and it undoubtedly contributed to concerns about corruption in the project, even though the real problem in this case, and in the project, was mostly carelessness.

† No, I didn't read the whole thing, and they were accused of trying to drown critics in paper.

temporarily into a pedestrian walkway, for example. (And then watch all the cyclists yell at the pedestrians in their space.)

Setting aside engineering and implementation challenges, there were plenty of other issues to consider. The proposal included not just Quayside but the wider Waterfront district, not part of the existing deal; this was a serious faux pas that clearly rattled Waterfront Toronto, ostensibly Sidewalk's partner. And, according to reports at the time, Alphabet had to promise to allow other broadband providers the chance to bid on rolling out the promised fibre network, and said it would build a system that would enable third-party sensor makers to be able to plug into the system, avoiding lock-in and a monopoly. And all the data that was collected would be managed by an independent trust.

Criticism came from many corners. One early critic was Jim Balsillie, best known as the one-time CEO of Research in Motion, aka RIM, the company that made BlackBerry a success and then an also-ran. Balsillie wasn't a fan of Silicon Valley types waltzing into Canada and taking business (and the eventual profit) that he believed should be kept local, calling the project 'a colonising experiment in surveillance capitalism attempting to bulldoze important urban, civic and political issues'. And then came the activists, led by local civics expert Bianca Wylie, who read an opinion piece by Doctoroff promoting the project and immediately took action, organising protests and making arguments to the government, media and even Sidewalk itself by attending their open-house events.

Sidewalk's own privacy team eventually quit the project. That included Ann Cavoukian,* the former privacy commissioner of Canada and creator of the Privacy by Design framework, which has been applied to the EU's GDPR. She

---

* An extremely Canadian reference: Ann Cavoukian is the sister of Raffi, the children's singer. His best-known works include 'Down by the Bay' and 'Baby Beluga'.

signed on to Sidewalk as a consultant, perhaps naively, but left after it became clear data would not be entirely stripped of personally identifiable information. 'I imagined us creating a Smart City of Privacy, as opposed to a Smart City of Surveillance,' she said in her resignation letter.

Even Waterfront Toronto's own board members quit, largely over Sidewalk's unrelenting push for more land but also citing concerns about process, patent ownership and legal challenges; others were fired. It's hard to imagine allowing people to manage a city district when they have so many difficulties organising themselves.

Sidewalk did take up offices near the site to show off a handful of potential technologies, O'Kane notes. That included those heated paving stones, which would not have survived Toronto's brutal winters, and 'rain jackets' for buildings, which were coverings that could stretch out like an awning, but which were positioned in such a way as to get in the way of those with reduced vision or blindness.

By the end of 2019, the two organisations finally agreed on a draft proposal, after two long years of work. But six months later, Sidewalk scrapped the project, with Doctoroff saying Quayside no longer made sense as COVID-19 hit and muddled up the works. 'Unprecedented economic uncertainty has set in around the world and in the Toronto real estate market, it has become too difficult to make the 12-acre project financially viable without sacrificing core parts of the plan we had developed together with Waterfront Toronto to build a truly inclusive, sustainable community,' he wrote in a Medium post.

At that time, Sidewalk had a host of other projects, but it's perhaps best associated with two in particular: Flow, a tool to battle traffic bottlenecks; and Link, which offered free local Wi-Fi in New York but surveilled streets with embedded cameras. Despite this, in 2020 Sidewalk itself was subsumed into Google.

While some of the technology suggested by Sidewalk sounds questionable – and stalkerish – the entire point of using a chunk of a city to build a smart test lab is to try new ideas. But it wasn't the tech that was the real problem with the Toronto project: it was Silicon Valley corporates clashing with democracy.

★ ★ ★

Building an entire smart city from the ground up is nigh-on impossible, as Songdo and Masdar show. Regenerating a district amid municipal bureaucracy and the pesky rules that comes with it is no walk in the carefully planned park, either. So can a smart city ever exist?

While on that Malaysia trip, we also visited Cyberjaya, half an hour from Kuala Lumpur. I had first heard of the city – Malaysia's own version of Silicon Valley – in an essay from Canadian journalist Chris Turner when building had begun in 2000. The *WIRED* story I wrote after my trip to Cyberjaya came nearly two decades later, yet a taxi driver told us he'd never been and never even heard of it. The police station's lawn was covered in dogs (which I described in my piece as 'feral', but I do admit in hindsight I have no idea of their provenance and perhaps they were pets, so it was indeed poor wording); and the streets, all named things like 'multimedia' and 'cyber', were mostly empty.*

Cyberjaya isn't empty; when I visited in 2016 there were some 800 companies located in the city, and 86,000 residents. It's no ghost town, but it's also no Silicon Valley – which was never a realistic goal anyway. (If we could plan a place like that, there'd be more than one.) During my visit, a local government representative told me the plan now was to turn

---

* Years later I was told my negative description of Cyberjaya – which honestly was fairly balanced if you read through to the end – apparently caused a diplomatic incident with trade mission officials. I'm sorry everyone; I really liked Malaysia, but there really were a lot of dogs on that lawn and not very many people in Cyberjaya. https://biturl.top/3MFbQj

Cyberjaya into a living lab for smart city technologies – yes, this again – be it digital signage or smart traffic lights or whatever else your startup can think up.

A local investor told me the government tried to encourage the right sort of companies to establish offices in Cyberjaya by offering them significant, long-running tax rebates. To get the cash, his company set up an office, but spent all their time in Kuala Lumpur. Because that's where the people are.

You can't have a smart city without people. And here's the thing: Kuala Lumpur is epic and exciting and brilliant. It *is* a smart city: it makes use of the various technologies for city management, has a smart city plan for future innovations, and most importantly, is a place where people want to be. Why would anyone try to build another Palo Alto when KL is right there? I can tell you which I'd rather spend a holiday in, and where I'd rather work.

We need more cities. And we need them to be sustainable and smart. But we need to stop playing *SimCity* and constructing cities top-down from fantasist master plans designed by famous architects who haven't built anything meaningful in years, and also stop building in technology that either doesn't yet exist or will be out of date by the time groundworks are completed.

Are we stopping with these shenanigans? No, we are not. Others have failed to learn these lessons, or believe they can overcome them. Toyota is building Woven City: a carbon-neutral, smart company town to act as a test lab for autonomous cars, with sensors everywhere, the data analysed by AI to make it all run better. Dutch architect Bjarke Ingels is designing a new city called Telosa in the American desert; Malaysia is ticking the sustainability box with BioDiverCity, set on artificial islands and using autonomous mass transport instead of cars; Foster + Partners is following up its Masdar City work with a plan for a city in India to be called Amaravati; and Chengdu Future City in China is also car-free, using automated mass transport with a master plan from Dutch architects OMA. Fancy architecture studios do well out of smart cities.

And then there's Saudi Arabia's NEOM. This massive project includes an octagonal floating city and two resorts, one in the mountains and the other an island in the Red Sea. But first up is the bewildering new mirror-walled city dubbed The Line, stretching 170km (105 miles), and 500m (1,600ft) tall, but just 200m (650ft) wide. There will be no cars and no roads, with transport running the length of The Line in 20 minutes, and people's necessities all within a five-minute walk. It's essentially a high-speed light-rail network with housing built around and atop it, surrounded by a mirrored facade.

When fully built – if it ever is – Saudi Arabia will house 9 million people in this building-width, very long city, giving it a population density more than double the existing leader, Manila in the Philippines. 'By leveraging AI technology, services are autonomous, saving you time and effort,' a marketing video explains, without explaining at all what that even means. 'Intelligent solutions create efficiency.' Another video promises AI will continuously learn what residents need to 'make life easier'.

I feel like it's boring to predict it won't be ever fully built and certainly never as promised; Songdo and Masdar sound positively reasonable in comparison.

What's intriguing about The Line is that, as with many of the most intensely future-focused ideas in this book, it has a historical precedent. Linear cities arranged around transport were first described in 1882 by Spanish city planner Arturo Soria y Mata, who wrote in Madrid's *El Progresso* that the most perfect city would be a 'single street unit 500 metres broad, extending if necessary from Cadiz to St. Petersburg, from Peking to Brussels'.

A town was built in this style, but only in limited trials as Soria y Mata lacked the funds to build the full rail network he planned, and eventually his Ciudad Lineal was swallowed up in the suburbs of Madrid.*

★ ★ ★

We don't need smart cities. We need good ones. We need liveable ones. We need sustainable ones. And we need them quickly. I mentioned earlier that the UN expects the ratio of people who live in cities to climb from half today to two-thirds by 2050. But that won't happen evenly, notes Ricky Burdett, a celebrated professor of urban studies at the London School of Economics where he is also Director of Urban Age, which runs global conferences on the future of cities. He told me in an interview for an article that this urbanisation would be distributed unequally, with most of it happening in Africa and Asia, noting that cities such as Lagos, Nigeria and Dhaka, Bangladesh are growing at a rate of hundreds of thousands of people each year. That's an average of 70 people every hour of the day, adding to existing burdens on everything from roads to sewers to hospitals.

Technology can help, but it doesn't hold all the answers – especially in places without reliable energy grids or budgets to fund such gadgetry. Burdett pointed to India's Smart Cities Mission, which had the admirable goal of providing

---

* Others followed suit, notes architecture critic Oliver Wainwright in the *Guardian*: https://biturl.top/U7byE3. Edgar Chambless planned a linear city in the 1930s in Virginia, US, dubbing it the 'roadtown – it was effectively an apartment block laid on its side, with 'noiseless' transportation making traversing the long city possible.' In the Soviet Union, Mikhail Okhitovich suggested a design for Magnitorgorsk, a city with 25km (15½-mile) strings of housing and reconfigurable pod houses – helpful in a divorce. French architect Le Corbusier offered up Plan Obus, a raised motorway with low-income housing built beneath it. Similar ideas were suggested in Japan, with architect Kenzo Tange suggesting a city across Tokyo Bay, and again back in the US, with a linear city linking Boston and Washington. None were ever built.

core infrastructure and improving quality of life, and planned to achieve that by adding smarts to 100 cities across the country. But as clever as automation and smart traffic lights are, he told me, these were cities that lacked sewers and houses and even toilets — shouldn't that come first?

There's a reason, after all, that so much smart city activity happens in wealthier nations: this stuff is expensive, and the companies selling it are expecting to take a cut, too. But the small changes that can revolutionise urban living are often simple. For an example of how to update a city for modern challenges, Burdett points to Colombia.

The city of Medellín faced an array of challenges, from gridlocked streets to drug cartels, addressing them with solutions centred on social equity. If that all sounds a bit academic or progressive to you, that's because it is, but put more simply, Medellín's administrators simply tried to make it a better place to live for the people who needed it the most. That meant that the poorest districts, known as barrios, where drug cartels held the most sway, saw the most investment, with new libraries, parks and schools, as well as a city-wide improvement to transport that included cable cars to end the isolation of the hillside barrios.

With its traffic-free squares and people-filled streets, Barcelona, Spain is also often touted as an ideal example of a smart city. This isn't because its gadgets and sensors and systems all work perfectly, but because it has rolled out technologies to solve human problems, be it when to pick up full bins or tackling questions of social justice. The city wasn't trying to be smart. And it wasn't just trying to balance a budget by slashing jobs. It was trying to be a nicer place for its citizens to live. Motivation matters and Barcelona is evidence of that. 'Smart cities are not an end in themselves but a means to an end,' argues Antoni Vives in his memoir of the project.

That's echoed by Josep-Ramon Ferrer, the director of Barcelona's smart city program. He advises in a paper basing a smart city centre not on what gadgets to choose and how to organise data, but on how to engage citizens, ensure appropriate governance and respond to local challenges – this isn't smart city management, this is city management. Core to the project was an online consultation platform, set up by Francesca Bria, then chief technology and digital innovation officer for the city, that let citizens share their views on government plans. In most cities, consultations are obscure, complex and hard to find. Barcelona made it easier to get feedback, not harder.

There's no question that tech can be used to improve city life. The card-based wave-and-pay ticketing system on London's public transport network is now more than 20 years old, but whenever I travel to another city I hate the friction of having to figure out how to buy a local bus ticket. This is a cheap and easy solution that makes it easier to pay for travel, though ensuring cash can still be used is crucial in some cities. Smart grids make it easier for energy companies to shift to renewable sources, while apps can allow a supplier to offer benefits for using energy at non-peak times to help balance demand.

But all of those are 'nice to haves' rather than requirements for a well-run city. What do people need in a city? Affordable homes. Clean air and water. Suitable mass transport. Safety and security. There are technologies that can help in each of those areas, but they're not the start or the end of any urban policy. Smart bus passes are a piece of sustainable transport, but you need buses and roads and perhaps bus lanes first. CCTV can help safety but hinder privacy. Sensors can track pollution, but surely we know that cars are what's polluting the air. Traffic filtering angers local drivers, governments aren't keen on the cost of river cleanup. And plenty of these areas intersect: you can improve buses all you want, but if they're unsafe, people won't use them.

We have simple, tech-free answers to most urban problems, though what's suitable depends on the local context. Tracking citizens and analysing their lives needs careful guardrails to protect against the worst excesses, but most of all it must only be done with a single purpose of improving the quality of life for everyone. And if a company's ideas threaten to take over the role of an elected democratic city in favour of privatised tech-driven solutions, we'd best show them the door.

Some technologies solve problems without impinging on privacy or risking it via hacks: coordinated traffic lights, bin notifications and energy reclamation are all useful without much in the way of a downside. Others have trade-offs. Smart bus passes might track users, but anonymous, aggregate data could be used to help plan routes and service frequency. Using Wi-Fi signals to track how many people use a public space could be useful for avoiding crowded transport, so long as identifying data isn't also hoovered up. Surveillance cameras make sense in a crime-ridden, extremely public space where there would be no assumption of privacy; add in AI-powered facial recognition, and it's a different story.

Few of us would be happy with cameras and sensors that watched our every move inside our own homes, even if it was ostensibly for our own health – but that's exactly what emergency health bracelets and the like do for elderly people or those with dementia. The surveillance gives them more control over their own lives, more safety and more privacy than living in a care home would, at least. Context matters, as does balance. That's why rolling out sensor-based technologies may not make sense across an entire city, but may be a perfect solution to address specific problems.

There's also the assumption that data is always accurate – which, as New York saw in the 1970s with fire station closures, is simply not the case – and belief in the unbiased scientific nature of algorithmic decision-making. Adam Greenfield notes in *Against the Smart City* that this is largely unremarked upon in smart city discussions, despite how easy it is to meddle with numbers: shift a sensor up or down a few metres and air pollution counts will differ; change the taxonomy for classifying crimes and a neighbourhood suddenly looks more (or less) safe.

Using tech to truly solve a problem is even better when it's an important issue like pollution, energy use or safety. Santander in Spain has connected parking lots, making it easier for drivers to find an empty spot, rather than drifting around wasting fuel and exacerbating pollution in search of a place to park. In the Netherlands, Utrecht has pollution-sniffing bikes to help monitor cycle routes so air quality can be improved. London has also sought to clean up its air, so sensors were installed on streetlights to monitor air quality. That's helpful, but do we really need sensors to tell us that the obvious answer is to reduce the use of cars, in particular the most polluting vehicles? Sensors may identify the worst air quality, giving an indication of where to start and what vehicles to target first, but that data is useless without political action.

Indeed, we should ditch the term 'smart cities'. The industry constantly changes what it means, and has realised it's no longer a selling point – after all, even Google's Sidewalk Labs said it practised 'urban innovation' – and city authorities seem all too aware of its waning ability to put a progressive sheen on mayors. And, with Songdo and Masdar City in mind, smart cities never seem to ever get built. Greenfield notes that delays and just-around-the-corner technologies are inherent to these types of plans and that smart cities will always be renamed and redefined, and therefore never actually completed.

If not smart cities, what shall we use in its place? Nothing. There's no such thing as a smart city. In her book *A City Is Not a Computer*, Shannon Mattern says that after writing about smart cities for a decade, she's had it with 'smartness'. She adds: 'I'm annoyed by its elasticity, ubiquity, and deceptiveness – and its sullying association with real estate development, "technosolutionism", and neoliberalism – so I plan to use the term as infrequently as possible.'

The term 'smartphone' once meant something, differentiating from the basic flip phones of years ago. But we now just call our iPhones and Androids what they are: phones. Yes, they're so much more. But slapping sensors on to lamp posts and using RFID cards for transport passes doesn't make a city 'smart' or 'intelligent' or 'advanced'; it's just one way to fight air pollution or encourage bus travel. There are other ways that don't use technology – modal filters using signs and barriers, free transport for young and elderly people – that are just as sensible. These are all just policies, to be used as civic leaders see fit.

We know what makes a city nice to live in. We know what people value. No one is asking for smart bins that ping a centralised dashboard, with data pulled into an AI decision-making system. But they are asking for clean streets and regular rubbish removal. If a sensor makes that happen without harming people through surveillance, that's fabulous. But first, make sure you've got an employee to go and empty that bin with a truck that won't get stuck in traffic.

CHAPTER NINE

Who Builds The Future

I'm sitting in a conference room at X labs, once known as Google X, awaiting Astro Teller and hoping he will skate in on rollerblades (he does!). It's a quirk of his – beyond the self-crafted name – that also lets the head of Google's innovation labs more easily navigate the modernist offices, formerly a small shopping mall in Mountain View.

Teller has run Google's X labs since 2010, overseeing projects from smart wearables like Glass to broadband-by-balloon Loon, ephemera of which fills a hallway to the left of reception that acts as the lab's museum of 'moonshots'. A rolling farming robot stands in one corner, while stuck to the wall near the front is a trio of fish models used in the early stages of the Tidal salmon farming project.

Teller's job is to systemise innovation, so I've come to X to ask: how do you build the future? With great difficulty – X labs has shuttered plenty of projects, and many more remain in development. Those don't count as failures: technology from one project carries over to another or sparks a fresh idea; Teller refers to it as 'moonshot compost'. Nothing is a failure if you learn from it, after all.

Consider driverless cars. Ernst Dickmanns' technology didn't spark a driverless revolution. But it did lead to improved safety equipment in cars. No one would say he wasted his time or the EU's money. Yet when we – and in particular I – talk about Google's driverless cars project, the focus tends to hover on the failings, the delays, the difficulties and the impossibilities. There are reasons for that. Google is a corporation, and Dickmanns is just one academic. Google already has tremendous reach into our lives; that is unlikely with Dickmanns. Google (and the rest) make huge (and incorrect) claims about when driverless cars arrive, breaking promises to remove steering wheels and brakes; Dickmanns' work is, if anything, grossly undervalued.

Deciding what projects to ditch is part of systemising innovation, argues Teller. 'It's really easy to create radical innovation, right? You just find a lot of very high-energy, crazy-sounding people, you give them a lot of money, a few of them will be right, you get radical innovation,' he says. 'But that is not an efficient process.'

But that is largely how the development of future tech has long worked, and continues to work. And to get that money, those high-energy, crazy-sounding people need to get attention. They do that through hype – VCs won't throw money at your idea if they haven't read about it on *TechCrunch* or in *WIRED*, after all. While X gets a steady supply of funds from Google, less lucky people in the industry, in particular startups, have to chase money or partnerships and so overstate their case. 'The way we fund innovation is mostly driven through that cycle,' Teller says, adding that would-be inventors aren't outright lying, but sort of secretly saying, *I wish I could do this, I've done a little of this, if you give me more money, I'll do more of this.* 'But then people get frustrated, because they thought it was kind of already done. As a society, that is not an efficient way to run innovation.'

Perhaps we need to learn patience, but so too do companies. Building the future takes time: Teller points to the iPhone, which had decades of predecessors from brick phones to Palm Pilots and BlackBerrys before it arrived. 'It was an overnight success 40 years in the making,' he says.

This is one reason why Apple doesn't make much of an appearance in this book. Future tech is by definition the ideas that haven't yet been realised – this is why this book is full of apparent failures. Apple's success is at least somewhat down to patience: it did not launch the first smartphone, tablet or smartwatch. There's no real value in being the first in anything, despite so many companies and inventors racing for that designation. Apple has long been rumoured to be building a driverless car, but refuses to admit any details about products in development, including their existence. (Whether that holds true for the foreseeable future remains to be seen, especially given Apple's entrance into the AR market.) This is helped by Apple's culture of secrecy and desire for product perfection, as well as its giant pile of cash, bolstered by tax avoidance techniques as well as commercial success. In short, Apple doesn't need hype to fund its R&D.

★ ★ ★

Perhaps I've asked Teller the wrong question. Forget *how* to build the future. Instead, *who* gets to build it? Answering that question also centres on money.

Many historical examples in this book were funded directly or indirectly by the military, and that naturally informs the sorts of projects that get funded – DARPA's first driverless car trial was set in a desert because that's where the US military was fighting (different desert, of course). Charles Rosen had to stretch the truth that Shakey could carry a gun to get his proposal approved, and this was for a valuable AI project that originated the algorithms that led NASA's rovers to wander Mars.

Recently, investment shifted to VCs and development consolidated in corporations, many grown, headquartered or funded from Silicon Valley, though increasingly China is pushing back on every front of tech development. But corporate funding means research and development is going to require a profit margin and business case, or to catch the imagination of someone very rich.

If you have money, like the billionaires whose names are peppered throughout this book, you can build what you like. If Elon Musk wants driverless software developed for his Teslas, all he has to do is pay for it. If Larry Page wants to play *SimCity* with real people and buildings, he can set up a company to have a go at it. And if Alfred Beach wants to build an air-blown train in a tunnel, well, he can (and did). Of course, reality still bites, even for billionaires: if driverless cars aren't technically possible, no amount of dollars will make them so; if a government or citizens don't want a bubble-domed city, Page's vision will stay on the page; and delayed by regulators and macroeconomics, Beach ran out of money and time to expand his pneumatic railway.

Still, proof of the consolidation of power and money in tech CEOs is evident with Musk, Page, Peter Thiel and the other billionaires whose names pop up repeatedly in any discussion of future technologies. They decide what gets funded, what gets built and how it is built. And that list, and most of the tech industry, is made up of white men, further entrenching their power – don't ignore that as mere 'identity politics', it's a simple fact that the tech sector has a particular makeup, and that tech is inherently about people. Read through any computing great's personal history – such as in Mountain View's Computer History Museum's excellent collection of Q&As – and it's a list of luminaries they met as teenagers, professors they were desperate to work with, and fellow students they were lucky enough to sit near. It matters if you're not in those rooms. And that naturally informs what projects get attention as well as how they are developed.

Let's start with the latter first. While working on a story for *WIRED* in the UK, I interviewed Tom Blomfield, one of the co-founders and then the CEO of a banking startup called Monzo. I looked forward to this interview because I use Monzo and love it. This banking app almost single-handedly pulled the old-fashioned British banking industry into the future, forcing the incumbents to be less terrible. Monzo has all sorts of lovely, tech-enabled tools – you can easily split payments, build a savings pot via automated roundups and manage budgets with whizzy charts. The company has also steadily worked on social features, such as helping unbanked people get accounts and letting customers ban themselves from dangerous spending like gambling. Fabulous stuff.

But the company also wasted time on silly projects. In the early days, the predominantly male development team wanted to track beer spending – across groceries, the pub and so on – mainly because young British men are amused by how much they drink. That team didn't, however, consider including a way to track spending on health, home or children, as none of those things were priorities to them. Blomfield said solving that requires diversity among staff, but Monzo also built a system for gathering suggestions from users, which results in ideas being added to a roadmap for keen Monzo fans to follow.

Consider a more significant example: AI. As we've seen, data sets reflect existing human biases, so models are embedded with racism and discrimination when trained. Multiple AI vision systems struggle to see black skin, for example. Imagine what happens when that algorithm is applied to driverless cars. White men working on these projects should spot these flaws, but given that these systems are released to the world without checking for such biases, it's clear that's not happening. Consider Elaine Herzberg, the woman killed by that Uber driverless car. The system – or rather, the people who built it – didn't expect a person to jaywalk in the dark across a four-lane road with a speed limit of 72km/h (45mph). The system did spot Herzberg, but was confused by her bicycle: it assumed a bike would be ridden forward at pace, as most of us would. But because Herzberg was homeless, that's not how she was using her bike, and instead she was pushing it slowly across the middle of the road with bags of belongings dangling from the handlebars. The car's AI system stumbled when faced with behaviour spurred by poverty.

Would it help to have more diversity in the companies developing future tech? Unquestionably. AI researchers such as Timnit Gebru and Joy Buolamwini, whom we met in an earlier chapter, use their time to watch for such flaws, aware of the impact as it reflects their lived experiences. They continue to do so, and thankfully, their warnings are heard – though perhaps less well heeded than the tech bro CEOs making headlines. But if we listened to Gebru and her co-authors, instead of hoovering up the world's digital data to shove it into a closed-box model that may or may not be better than previous iterations, we might find it's worth trying to build more intelligent models that thoughtfully use better-quality data.

If this approach works (and it may not), generative AI would use less energy and emit fewer emissions, cost less money, and therefore be more widely accessible, and have better outputs without as many problems with bias and the like. Why build bigger when you can build better? After a while, we will run out of data and be, once again, restricted by computing power. The organisation that takes a wiser approach now may fall behind in the short term, but leap ahead when others hit these inherent, looming restrictions.

By limiting who gets to work on these projects, we limit the ability for these technologies to succeed. Would hiring women, people of different ethnicities, or from varied economic backgrounds make flying cars viable or solve the challenges facing driverless cars? Maybe the idea that makes driverless cars truly smart is sitting in a young woman's brain in Nigeria or Argentina or Malaysia. Perhaps not – physics is the same everywhere, after all – but getting her opinion might help avoid some of the harms along the way, and spark some solutions that 70-plus years of old-school, white, male academia hasn't managed to come up with on its own. I'm not saying to leave those dudes out – please do include them. But perhaps we can all have a say in how the future is built.

And people who live differently from us might have entirely different issues which need more urgent solutions than the problems prioritised in the West. For example, Kenya had mobile banking via the M-PESA text messaging system back in 2005, a clever tech solution to help the millions of unbanked people in the region – that was actually inspired by their own kludge of using mobile airtime credits to stand in for cash. At the time, it still took a UK bank three days to transfer money between accounts.

But not all problems have tech answers. If we leave it up to engineers, Silicon Valley billionaires or the like, they're going to keep looking to technology for answers. It's not a surprise that an internet company believes more internet is the key to development and fighting social ills. It's why a technologist like Sebastian Thrun thinks driverless cars are the right tool to end road accidents. And it's why aerospace engineers look to redesign air mobility to address pollution concerns rather than try to reduce flight.

It's Maslow's law of the instrument: 'If the only tool you have is a hammer, it is tempting to treat everything as if it were a nail.'* We work with the tools we have, the knowledge we already know, to solve a problem. This, in a way, is the promise of AI: if a system can look at a problem from every angle, with data from many sources, perhaps we can finally get better solutions. But it's also the downfall of AI: right now it's trained on poorly curated nonsense gobbled up from the gibberish on the wider web. Remember expert systems? The whole idea was to curate knowledge and automate its application carefully. We have answers to many of the problems right in front of our faces, if we only have the memory to look back now and then and consider the experiences of others.

Flying cars have been suggested for decades as a way to rise above traffic. Why not solve traffic? Driverless vehicles are mooted as a way to reduce emissions and improve safety. Yet they add cars to the road, block emergency vehicles and, as

* As with any well-known quote, there's discrepancy over who actually said it first, but psychologist Abraham Maslow did actually write this in a book in 1966.

Thrun himself admits, have yet to save a life. Some problems can't be fixed by technology, Astro Teller admits. One that haunts him is global education. 'I still want to have a moonshot or two here in education, I feel like it's one of the world's biggest problems … but nothing stuck. Maybe technology's just not going to be able to solve that.'

If I were handed Teller's job at X labs, a few things would happen. I would collapse under the stress and pressure of it all and be humiliated by my lack of engineering knowledge and basic maths skills. But if I could get anyone to listen to me, I know what projects I would focus on. Feeding the planet without burning it to the ground. Climate crisis avoidance and mitigation techniques. Solutions for the havoc wreaked by the technology industry, such as developing better data sets and methods to assess them, using AI to improve AI, and developing systems to answer serious questions rather than act as a chatbot for bored office workers.

Can you please, I ask Teller, set the minds of the brilliant people working in this warehouse-chic lab to solve today's biggest challenges, rather than just getting broadband coverage a bit better or making cars driverless? He agrees and says half of the current projects at X are now climate-themed. But it's a shame so much time was wasted getting to this point. And, this being a corporate-funded lab, money matters. 'We're going to have to find a way for doing the right thing to be cheaper than doing the wrong thing, and that's only going to happen because the technology innovation cycle gets us there,' says Teller. 'On climate change, I think it is both necessary and coming fairly quickly.'

Many people making decisions about future technologies have a lot of power, and their considerable influence is not addressing the challenges that impact all of us but those that they value personally. Instead of tackling global concerns, Amazon's Jeff Bezos wants to colonise space, as does Elon Musk, who has so much influence thanks to his satellite network that the US military is reportedly worried about his foray into the war in Ukraine. And that's before we get into the influence of Twitter, now named X, on politics. Other tech leaders have bonkers plans to escape government via seasteading*, have built end-of-the-world bunkers in case they accidentally spark a dystopia, and are actively trying to live forever using the blood of the young – all three of these apply to Peter Thiel, co-founder of PayPal and venture capitalist. He gets to decide what gets built.

★ ★ ★

How do we retake control of the future? Take their money and make what we want? Okay, okay fine, that's not going to happen, though you have to admit it has appeal. But we can take lessons from the past, from what has worked. Sidewalk Labs blamed the COVID-19 pandemic for making the Waterfront Toronto project unviable, but much of the credit should be handed to local activists such as Bianca Wylie. She didn't like the future of the city as planned by Google – privatised, tech solutionism with automated transport – so she stopped it. Similar is happening in San Francisco with driverless cars, with activists taking on Google's Waymo and GM's Cruise with nothing more than a bit of free time and traffic cones. And Timnit Gebru and Margaret Mitchell may have lost their jobs at Google, but they are winning plenty of headlines and, perhaps, the argument regarding AI. People can push back against billionaires' bad ideas.

Activism alone won't save us. We need regulators who understand how technology works to see its flaws, minimise harms and not be fooled by Silicon Valley lobbying or marketing – and voters literate enough to hold them

* Seasteading is when you build a place to live in international waters in order to escape government interference (you can't just park a boat out there, to be clear). Thiel is one of the cofounders of The Seasteading Institute.

accountable if they sell out. If eVTOLs aren't safe or suitable for urban environments, they shouldn't be allowed to hover over cities. If AI is harming people in the here and now, hold someone responsible in the here and now.

And how do we fund projects that benefit us all? Think back to the World's Fair in 1939: Norman Bel Geddes argued that the nation needed driverless cars, but also motorways – and that the latter should be paid for by the government. With the benefit of hindsight, he was wrong on how the money should be spent, but he successfully convinced politicians to shell out on building a better future. No one believes bureaucracy is the right path to innovation, but large government organisations certainly worked well with the internet *and* the web, with DARPA's driverless and robotics kickstarters, and even with flying cars – there is no aerospace company on the planet that doesn't go on and on about how closely it works with regulators.

This is all getting quite serious, and not every technology in this book is threatening human existence, of course. There's nothing to fear with virtual or augmented reality beyond a bit of nausea and ugly glasses. And hyperloops don't exist, so what's the problem? There is an opportunity cost, as we risk failing to invest in the right things – say, green transport – when we focus on nonexistent options, like hyperloop. The same holds for robotics. If enough people believe humanoid robot helpers will solve the looming care crisis, we'll fail to address it in good time. Not all problems can be solved by technology. And that's especially true if you can't manage to make it work.

But the point of this book isn't to laugh at the mistakes and failings of engineers, as hilarious as some of them are – flying cars crashing from mistaken fuel gauges, nausea-inducing red-only VR tripods, robots tripping over themselves. Building anything is hard. And science for the sake of science is worth celebrating. Ivan Sutherland made the first head-mounted augmented reality display, just to see how it would work, before setting aside the technology and spending his time solving other problems. We should celebrate such engineering.

If this time-travelling tour of future tech helps you understand anything about this bonkers industry, I hope it's this: none of this technology is inevitable; turning science into reality is sometimes impossible; and we get to decide what happens next. So do you still want flying cars, hyperloops, and driverless vehicles? Or perhaps … something better.

Acknowledgements

I wrote too many words – to the surprise of no editor I've ever worked with – so my thank yous have been limited to a couple of pages, which is not enough but I'll use it as an excuse in case I forget anyone. Let's start by thanking my fabulous agent, Donald Winchester, for all the help and guidance, and Ciara McEllin the rest of Watson, Little. Next, thanks to everyone at Bloomsbury, particularly Jim Martin for taking on this project and Sarah Lambert for making it happen.

Something that's always surprised me as a journalist is how generous interviewees are with their time – even when they have no product to pitch. I am very grateful to everyone who took the time to speak to me, especially Reinhold Behringer, Bob Sproull, Tony Claydon, Joan Lobban, Daniel Ermisse, and the many PRs who helped me set up interviews. I'd have been lost without the assistance of staff at the British Library, Stanford Archives, and the Science Museum's Dana Research Centre and Library.

Thanks to my parents, Al and Monika, for their support, and to my mother-in-law Carol for that and the babysitting too – oh, I know it's no trouble, but thank you all the same. Michelle, Melanie, Anna, Maresa, Shannon, Ivy, and Harriet, thanks for checking in on me; Davina, Jen and Dean, thanks for listening to my rants about driverless cars – and esports. Tina, this book is better because of our conversations. Mustafa, thank you for the photo and the chats. And because seeing your name in print is fun, shout out to my nieces and nephews: Maella, Evia, Jack, Sofia and Kayson.

Writing this meant testing the patience of my editors in my day job, so thanks for tolerating my stressed vibe and loose interpretation of deadlines. Tim Danton, you're an excellent boss and an even better person – beers soon, I'm buying. And thank you to *PC Pro* readers; you're a lovely audience.

None of this would have happened without my husband, Michael, who gave me the confidence to pitch this idea, created the space to make it happen, cheerfully tolerated me running into the room with another "fun fact" from my research, and propped me up through my weekly meltdowns at 11am every Tuesday – and then edited, proofed and fact-checked every word. You're the best. I am so lucky.

Sitting at my desk for long hours was less lonely because of my fluffy little buddy Willow, even if all you did was sleep on my pillow all day; you were a good girl. And there is no better, more welcome interruption than my delightful daughter Eliza running up behind my chair demanding hugs and giraffe videos. I may not need to know what sound a giraffe makes for this book, but I appreciate your love for fun facts.

Bibliography

Intro

Crevier, Daniel. *AI: The Tumultuous History of the Search for Artificial Intelligence*. New York: Harper Collins, 1993.

Kobie, Nicole. 'No, driverless cars will not be racing around UK roads this year.' *WIRED*, February 7, 2019. https://biturl.top/raaqU3

'Arthur Rock: Silicon Valley's Unmoved Mover.' The Generalist. https://biturl.top/MramMn Accessed August 23, 2023.

Topham, Gwyn. 'First hands-free self-driving system approved for British motorways.' *Guardian*, April 15, 2023. https://biturl.top/6F36ny

UK Government press release. 'Government moves forward on advanced trials for self-driving vehicles.' Published February 6, 2019. https://biturl.top/ErEbAr

Driverless Cars

Ackerman, Evan. 'Carnegie Mellon Solves 12-Year-Old DARPA Grand Challenge Mystery.' *IEEE Spectrum*, October 19, 2017. https://biturl.top/jEvUBz

Ackerman, Evan. 'The electronic highway: How 1960s visionaries presaged today's autonomous vehicle.' *IEEE Spectrum*, August 2, 2016. https://biturl.top/BbeU73

Bel Geddes, Norman. *Magic Motorways*. New York: Random House, 1940.

Bhuiyan, Johana. 'Self-driving cars are here and they're watching you.' *Guardian*, July 4, 2023. https://biturl.top/zueIVv

Boffey, Daniel. 'World's first electrified road for charging vehicles opens in Sweden.' *Guardian*, April 12, 2018. https://biturl.top/fmYZ73

Bursa, Mark. 'TfL claims 20mph speed limit has cut accidents and fatalities in London.' *Professional Driver*, February 15, 2023. https://biturl.top/3ym2Qz

Cardew, K. H. F. 'The Automatic Steering of Vehicles – an experimental system fitted to a Citroen DS 19 Car.' Road Research Laboratory (Department of Transport), 1970. (From Science Museum archives.)

Dave, Paresh. 'Dashcam Footage Shows Driverless Cars Clogging San Francisco.' *WIRED*, April 10, 2023. https://biturl.top/QnMFVf

Davies, Alex. *Driven: The Race to Create the Autonomous Car.* New York: Simon & Schuster, 2021.

Delcker, Janosch. 'The man who invented the self-driving car (in 1986).' Politico, July 19, 2018. https://biturl.top/Iveui2

Dickmanns, E. D. 'The development of machine vision for road vehicles in the last decade.' Intelligent Vehicle Symposium, 2002. IEEE, Versailles, France, 2002, pp. 268–281, vol.1. doi: 10.1109/IVS.2002.1187962

Ernst Dickmanns, an oral history conducted in 2010 by Peter Asaro, Indiana University, Bloomington Indiana, for Indiana University and the IEEE. https://biturl.top/Z7Vziq

Etherington, Darrell and Conger, Kate. 'Uber's Anthony Levandowski invokes Fifth Amendment rights in Waymo suit.' TechCrunch, March 31, 2017. https://biturl.top/RNb6ja

Fairfield, Nathaniel (Waymo). 'On the road with self-driving car user number one.' Medium/Waymo, December 13, 2016. https://biturl.top/73yqEr

Fenton, R. E. and Olson, K. W. 'The electronic highway.' *IEEE Spectrum*, vol. 6, no. 7, pp. 60–66, July 1969. doi: 10.1109/MSPEC.1969.5213898

Griffiths, Hugo and Saarinen, Martin. 'Wireless electric car charging: is EV charging without cables the future?' *Auto Express*, January 17, 2020. https://biturl.top/Zba22e

Harris, Mark. 'How a robot lover pioneered the driverless car, and why he's selling his latest to Uber.' *Guardian*, August 19, 2016. https://biturl.top/UNbUJ3

Harris, Mark. 'How Otto Defied Nevada and Scored a $680 Million Payout from Uber.' *WIRED*, November 28, 2016. https://biturl.top/mIZZ7j

Jochem, Todd; Pomerleau, Dean; Kumar, Bala; Armstrong, Jeremy. 'PANS: a portable navigation platform.' Intelligent Vehicle Symposium, IEEE, 1995. https://biturl.top/zEBFNn

Kerry, Cameron F. and Karsten, Jack. 'Gauging investment in self-driving cars.' Brookings Institution, October 16, 2017. https://biturl.top/JF3qEf

Korosec, Kirsten and Harris, Mark. 'Anthony Levandowski sentenced to 18 months in prison as new $4B lawsuit against Uber is filed.' TechCrunch, August 4, 2020. https://biturl.top/BbyABb

Lockton, Georgina. 'The driverless car in 1960s Britain.' Institute of Historical Research Seminar Series: Transport & Mobility History, January 15, 2021. https://biturl.top/AFvAFr

Markoff, John. 'Google Cars Drive Themselves, in Traffic.' *New York Times*, October 9, 2010. https://biturl.top/UNbUJ3

Marshall, Aairan. 'An Autonomous Car Blocked a Fire Truck Responding to an Emergency.' *WIRED*, May 27, 2022. https://biturl.top/FrmeAf

McCullagh, Declan. 'Robotic Prius takes itself for a spin around SF.' CNET, September 26, 2008. https://biturl.top/by6NFv

Nichols, Shaun. 'Uber responds to Waymo: We don't even use that tech you say we stole.' The Register, April 8, 2017. https://biturl.top/ZrmIzm

NOVA. Season 33, Episode 9, 'The Great Race,' directed by David Cohen, aired February 22, 1995 on PBS. https://biturl.top/YrQFV3

Null, Schuyler; Bray Sharpin, Anna; Cunha Tanscheit, Paula. '6 Road Design Changes That Can Save Lives.' World Resources Institute, October 3, 2018. https://biturl.top/R7RzIv

Penoyre, Slade. 'A robot in the driver's seat.' *New Scientist and Science Journal*, May 13, 1971. https://biturl.top/EZBRfu

Penoyre, Slade. 'Automated Road Vehicles.' Transport and Road Research Laboratory, March 23, 1972. (Via Science Museum archives)

'Preliminary Report: Highway, HWY18MH010.' National Transport Safety Board, November 7, 2019. https://biturl.top/BNveme

Quigg, Doc. 'Reporter Rides Driverless Car.' *Press-Courier*, June 7, 1960. Accessed via https://biturl.top/iiQBB3

'Highway of the Future.' *RCA Electronic Age*, January 1958. Accessed via https://biturl.top/bUjaye

'Vehicle Automation: Principles of Automatic Systems.' Road Research Laboratory Leaflet 265. (Via Science Museum archives)

'Vehicle Automation: Brake and Throttle Control.' Road Research Laboratory Leaflet 260. (Via Science Museum archives)

'The RRL Electro-Pneumatic Servo Unit for Throttle Control on Petrol Engine Vehicles.' Road Research Laboratory Leaflet 261, March 1971. (Via Science Museum archives)

https://www.reuters.com/article/us-uber-selfdriving-sensors-insight/ubers-use-of-fewer-safety-sensors-prompts-questions-after-arizona-crash-idUSKBN1H337Q

Roth, Emma. 'Here's what happens when cops pull over a driverless Cruise vehicle.' The Verge, April 12, 2022. https://biturl.top/Un2Qrm

Shaban, Bigad; Horn, Michael; Carroll, Jeremy. 'San Francisco city attorney files motion to pump the brakes on driverless cars.' NBC Bay Area, August 17, 2023.

Shepardson, David. 'Fatal Tesla Autopilot crash driver had hands off wheel: U.S. agency.' Reuters, June 7, 2018. https://biturl.top/6BrMj2

Siddiqui, Faiz and Merrill, Jeremy B. '17 fatalities, 736 crashes: the shocking toll of Tesla's Autopilot.' *Washington Post*, June 10, 2023. https://biturl.top/2qU3eq Accessed August 19, 2023.

Snyder, Jon. '1939's "World of Tomorrow" Shaped Our Today.' *WIRED*, April 29, 2010. https://biturl.top/zyYzay

Somerville, Heather; Lienert, Paul; Sage, Alexandria. 'Uber's use of fewer safety sensors prompts questions after Arizona crash.' Reuters, March 28, 2018. https://biturl.top/JJfe6z

'How much does a mile of road actually cost?' Strong Towns. https://biturl.top/qE7Fze Accessed October 12, 2023.

Stumpf, Rob. 'Early driverless cars used underground magnets to test in California 25 years ago.' The Drive, July 18, 2023. https://biturl.top/nMJvUn

'Full self-driving capability subscriptions.' Tesla. https://biturl.top/6bmEfe Accessed August 18, 2023.

@that_mc. 'When an autonomous vehicle causes a collision, it wasn't tired, or intoxicated, it didn't get distracted or try to get away with something it knew better than to do. It "believed" it was driving correctly. They don't work as advertised, and they shouldn't be on the road.' Twitter, March 24, 2023. 4.53pm. https://biturl.top/J7J3uu

Valinsky, Jordan. '"Complete meltdown": Driverless cars in San Francisco stall causing a traffic jam.' CNN, August 14, 2023. https://biturl.top/IjQrii

Vance, Ashlee. 'DARPA's Grand Challenge proves to be too grand.' The Register, March 13, 2004. https://biturl.top/MzYV3m

Wakabayashi, Daisuke. 'Uber and Waymo Settle Trade Secrets Suit Over Driverless Cars.' *New York Times*, February 9, 2018. https://biturl.top/BRjUf2

Artificial Intelligence

Angwin, Julia; Larson, Jeff; Mattu, Surya; Kirchner, Lauren. 'Machine Bias.' ProPublica, May 23, 2016. https://biturl.top/2UZ36j

Lighthill Controversy Debate, BBC, 1973. https://biturl.top/N7nIje

Bender, Emily M.; Gebru, Timnit; McMillan-Major, Angelina; Mitchell, Margaret. 'Statement from the listed authors of Stochastic Parrots on the "AI pause" letter.' DAIR Institute, March 31, 2023. https://biturl.top/neqMJr

Bender, Emily M.; Gebru, Timnit; McMillan-Major, Angelina; Shmitchell, Shmargaret. 'On the Dangers of Stochastic Parrots: Can Language Models Be Too Big?' *FAccT '21: Proceedings of the 2021 ACM Conference on Fairness, Accountability, and Transparency, March 2021*, pp. 610–623. https://doi.org/10.1145/3442188.3445922

Bernhardt, Chris. *Turing's Vision: The Birth of Computer Science.* Cambridge, London: The MIT Press, 2016.

Brooks, Rodney. 'How Claude Shannon Helped Kick-start Machine Learning.' IEEE Spectrum, January 25, 2022. https://biturl.top/Qfauea

Brooks, Rodney. 'Intelligence Without Representation.' *Artificial Intelligence*, volume 47, issues 1–3, January 1991, pp. 139–159. https://doi.org/10.1016/0004-3702(91)90053-M

Bubeck, Sebastien; Chandrasekaran, Varun; Eldan, Ronen; et al. 'Sparks of Artificial General Intelligence: Early experiments with GPT-4.' Microsoft Research, March 22, 2023. https://biturl.top/Nnqy2e

Burgess, Matt. *Artificial Intelligence: How Machine Learning Will Shape the Next Decade.* London: Penguin Random House, 2021.

'Artificial Intelligence Startups see 302% Funding Jump in 2014.' CB Insights. https://biturl.top/JzINne Accessed August 29, 2023.

Crevier, Daniel. *AI: The Tumultuous History of the Search for Artificial Intelligence.* New York: Harper Collins, 1993.

'Pause Giant AI Experiments: An Open Letter.' Future of Life Institute. https://biturl.top/ba6f2e Accessed August 27, 2023.

Gershgorn, Dave. 'The data that transformed AI research – and possibly the world.' Quartz, July 26, 2017. https://biturl.top/22iEbe

Godfrey-Smith, Peter. 'Is Longtermism Such a Big Deal?' *Foreign Policy*, November 12, 2022. https://biturl.top/fuYvEf

'What are neural networks.' IBM. https://biturl.top/6Freae Accessed October 18, 2023.

John McCarthy, an oral history conducted in 2011 by Peter Asaro with Selma Šabanovic, Indiana University, Bloomington Indiana, for Indiana University and the IEEE. https://biturl.top/bYNBBn

Johnson, Khari. 'How Wrongful Arrests Based on AI Derailed 3 Men's Lives.' *WIRED*, March 7, 2022. https://biturl.top/EnM3yq

Kobie, Nicole. 'NVIDIA and the battle for the future of AI chips.' *WIRED*, July 17, 2021. https://biturl.top/MV3yay

Kobie, Nicole. 'Facial recognition for bears (and other ways to use the technology for good).' *PC Pro*, February 2021. https://biturl.top/RNz2yq

Kobie, Nicole. 'Watching the oceans using off-the-shelf tech.' *PC Pro*, September 2021. https://biturl.top/bYnUze

Kobie, Nicole. 'We asked an AI to write the Queen's Christmas speech.' *WIRED*, December 24, 2020. https://biturl.top/n6NbQv

Lewis, Tanya. 'One of the Biggest Problems in Biology Has Finally Been Solved.' *Scientific American*, October 31, 2022. https://biturl.top/Arq6Bf

Markoff, John. 'Computer Eyesight gets a lot more Accurate.' *New York Times*, August 14, 2014. https://biturl.top/fayuea

McClelland, James L. and Rumelhart, David E. *Parallel Distributed Processing: Explorations in the Microstructure of Cognition*. Cambridge, Massachusetts, MIT Press: 1987.

McCulloch, Warren S. and Pitts, Walter. 'A Logical Calculus of the Ideas Immanent in Nervous Activity'. *Bulletin of Mathematical Biophysics*, vol. 5, pp. 115–133 (1943). https://biturl.top/v26nEv

McKie, Robin. '"Only AI made it possible": scientists hail breakthrough in tracking British wildlife.' *Guardian*, August 13, 2023. https://biturl.top/iyuQ3a

Meade, Amanda. 'News Corp using AI to produce 3,000 Australian local news stories a week.' *Guardian*, July 31, 2023. https://biturl.top/m2AnEr

'A Concise History of Neural Networks.' Towards Data Science/ Medium, August 13, 2016. https://biturl.top/Q7B77v

Metz, Cade. '"The Godfather of A.I." Leaves Google and Warns of Danger Ahead.' *New York Times*, May 4, 2023. https://biturl.top/Q7bANb.

Metz, Cade. *Genius Makers: The Mavericks Who Brought AI to Google, Facebook and the World*. London: Penguin Random House, 2021.

Metz, Cade. 'Genius Makers: The Mavericks Who Brought AI to Google, Facebook and the World.' New York: Random House Business, 2021.

Milmo, Dan. 'Google chief warns AI could be harmful if deployed wrongly.' *Guardian*, April 17, 2023. https://biturl.top/IJjuqi

Minsky, Marvin and Papert, Seymour. *Perceptrons: An Introduction to Computational Geometry*. Cambridge, Massachusetts, MIT Press, 1969. https://biturl.top/e6Fv6z

Musk, Elon (@elonmusk). 'Worth reading Superintelligence by Bostrom. We need to be super careful with AI. Potentially more dangerous than nukes.' Twitter, August 3, 2014, 3.33am. https://biturl.top/QfUjYr

'New Navy Device Learns by Doing.' *New York Times*, July 8, 1958. https://biturl.top/IzQBzq

Ongweso, Edward, Jr. 'OK, WTF Is "Longtermism", the Tech Elite Ideology That Led to the FTX Collapse?' *Vice*, November 23, 22. https://biturl.top/UZnIna

'Better language models and their implications.' OpenAI. https://biturl.top/nEjq2i Accessed October 10, 2023.

Roose, Kevin. 'How ChatGPT Kicked Off an AI Arms Race.' *New York Times*, February 3, 2023. https://biturl.top/qEBJny

Rosenberg, Chuck. 'Improving Photo Search: A Step Across the Semantic Gap.' Google Research Blog, June 12, 2013. https://biturl.top/6ZniMv

Rumelhart, D.; Hinton, G.; Williams, R. 'Learning representations by back-propagating errors.' *Nature* 323, 533–536 (1986). https://doi.org/10.1038/323533a0

Russon, Mary-Ann. 'OpenAI boss Sam Altman tells congress he fears AI is "harm" to the world.' *Evening Standard*, May 17, 2023. https://biturl.top/EZ7nQf

Sack, Harald. 'Marvin Minsky and Artificial Neural Networks.' SciHiBlog, August 9, 2020. https://biturl.top/aEjqia

Shane, Janelle. *You Look Like a Thing and I Love You: How Artificial Intelligence Works and Why It's Making the World a Weirder Place.* London: Headline Publishing Group, 2019.

Turing, A. M. 'Computing Machinery and Intelligence.' *Mind*, volume LIX, issue 236, October 1950, pp. 433–460. https://doi.org/10.1093/mind/LIX.236.433

Turing, A. M. 'On Computable Numbers, with an Application to the Entscheidungsproblem.' *Proceedings of the London Mathematical Society*, volume s2-42, issue 1, 1937, pp. 230–265. https://doi.org/10.1112/plms/s2-42.1.230

Wooldridge, Michael. *The Road to Conscious Machines: The Story of AI.* London: Pelican, 2021.

Robots

Ackerman, Evan. 'Your robotic avatar is almost ready.' *IEEE Spectrum*, April 16, 2023. https://biturl.top/1VfI7b

Buchanan, Wyatt. 'Charles Rosen – expert on robots, co-founder of winery.' SFGate, December 20, 2002. https://biturl.top/nq6Vvq

CNET Highlights. 'Elon Musk Reveals Tesla Bot.' YouTube, August 20, 2021. https://biturl.top/E3mQfi

Darrach, Brad. 'Meet Shaky, the first electronic person.' *Life*, November 20, 1970.

Dunn, Angela Fox. Correspondence and article. October 6, 1980. Department of Special Collections, Stanford University Archives: M1275, Charles Rosen Papers, Box 2.

Edwards, David. 'UBTECH Robotics unveils latest version of Walker X humanoid.' *Robotics and Automation*, August 9, 2021. https://biturl.top/bEnEr2

'Eric, the robot, at Kennards.' *Croydon Times*, February 15, 1936.

'1928 – Eric Robot – Capt. Richards & A.H. Reffel.' Cybernetic Zoo. https://biturl.top/eiEnyq Accessed 25 July 2023.

'Rebuilding Eric: The UK's First Robot.' Kickstarter. https://biturl.top/3m2mUb Accessed 10 October, 2023.

'Eric the Erudite.' *Lincolnshire Echo*, October 10, 1933.

'Eric the robot again: Mechanical Man marvel at Model Exhibition in London.' *Sunday Mirror*, September 8, 1929.

Gale, Alastair and Mochizuki, Takashi. 'Robot Hotel Loses Love for Robots.' *Wall Street Journal*, January 14, 2019. https://biturl.top/NB3Azq

Guizzo, Erico. 'How Aldebaran Robotics built its friendly humanoid robot, Pepper.' *IEEE Spectrum*, December 26, 2014. https://biturl.top/3I3EFf

Heater, Brian. 'Figure's Humanoid robot takes its first steps.' TechCrunch, May 18, 2023. https://biturl.top/VzYJri

Hiler, Katie. 'The Automatons of Yesteryear.' *New York Times*, October 28, 2013. https://biturl.top/jyaqAf

'Asimo History.' Honda. https://biturl.top/FjyER3 Accessed August 29, 2023.

'Robot development history.' Honda. https://biturl.top/yeaiee Accessed July 1, 2023.

'Honda's Asimo conducts Detroit Symphony Orchestra.' YouTube, May 15, 2018. https://biturl.top/rINZJf

Inada, Miho. 'Humanoid Robot Keeps Getting Fired From His Jobs.' *Wall Street Journal*, July 13, 2021. https://biturl.top/AjqAVn

Kuipers, Benjamin; Feigenbaum, Edward A.; Hart, Peter E.; Nilsson, Nils J. 'Shakey: From Conception to History.' *AI Magazine*, 2017. https://biturl.top/aEfA3a

Markoff, John. *Machines of Loving Grace: The Quest for Common Ground Between Humans and Robots*. New York: HarperCollins, 2015.

Marsh, Allison. 'In 1961, the First Robot Arm Punched In.' *IEEE Spectrum*, August 30, 2022. https://biturl.top/vUBZZj

Mickle, Paul. 'A peep into the automated future.' *Trentonian*, 1961. https://biturl.top/JzQnE3

Mims, Christopher. 'Why Japanese Love Robots (And Americans Fear Them).' *MIT Technology Review*, October 12, 2010. https://biturl.top/MZfuim

'Robots In Progress.' MIT Leg Laboratory. https://biturl.top/zUZvaa Accessed October 14, 2023.

'Oral History of Peter Hart, part 1.' Interviewed by David Brock, May 16, 2018. Computer History Museum. https://biturl.top/JjArqq

'QI ASIMO.' YouTube, January 7, 2014. https://biturl.top/N7fIri

Raibert, Marc. 'Making Robots Smarter in Body and Mind.' ICRA London Keynote, June 1, 2023.

Russell, Ben (editor). *Robots: The 500-Year Quest to Make Machines Human*. London: Scala Arts & Heritage Publishers, 2017.

SRI International. '75 Years of Innovation: Shakey the Robot.' Medium, April 2, 2020. https://biturl.top/vIvuue

SRI International. 'Shakey the Robot: the first robot to embody Artificial Intelligence.' YouTube, February 14, 2017. https://biturl.top/BruqQb

Takenaka, Kiyoshi and Sakoda, Mayu. 'Sony says it has technology for humanoid robots, needs to find usage.' Reuters, December 6, 2022. https://biturl.top/f6JNv2

Unimate sales brochures, in the digital collection of the Henry Ford Museum. https://biturl.top/2uqAj2 and https://biturl.top/ERRfEv and https://biturl.top/iaa2Av

'Robot Takes a Tumble at Chicago Supply Chain Exhibit.' Yahoo News. April 10, 2023. https://biturl.top/qmQfaa

Young, Liz. 'Robots Are Looking to Bring a Human Touch to Warehouses.' *Wall Street Journal*, June 12, 2023. https://biturl.top/eqay2i

Zaun, Todd. 'Now, a Robot That Toots Its Own Horn.' *New York Times*, March 15, 2014.

Zaun, Todd. 'Why did Honda Build a Humanoid Robot that Meets with the Vatican's Approval?' *Wall Street Journal*, September 4, 2001. https://biturl.top/qEb2eu

Augmented reality

ACM SIGCHI. 'VIEW: The Ames Virtual Environment Workstation.' YouTube, February 9, 2021. https://biturl.top/R3QRna

Albergotti, Reed. 'The Reality Behind Magic Leap.' The Information, December 8, 2016. https://biturl.top/vYJFJn

Antonoff, Michael. 'Living in a virtual world.' *Popular Science*, June 1993, p. 82. https://biturl.top/r6Zzuu

Barras, Colin. 'How virtual reality overcame its "puke problem".' *BBC Future*, March 27, 2014. https://biturl.top/qEJvua

Capaccio, Anthony. 'Microsoft's Army Goggles Left US Soldiers with Nausea, Headaches in Test.' Bloomberg, October 13, 2022. https://biturl.top/zQ7Bby

Chafkin, Max. 'Why Facebook's $2 Billion Bet on Oculus Rift Might One Day Connect Everyone on Earth.' *Vanity Fair*, October 2015. https://biturl.top/ZV7viu

Chen, Brian X. 'A First Try of Apple's $3,500 Vision Pro Headset.' *New York Times*, June 6, 2023. https://biturl.top/YvUnuy

Computing Machinery's (ACM) Special Interest Group on Computer Graphics and Interactive Techniques. 'VR@50 panel discussion.' YouTube, August 13, 2018. https://biturl.top/FRvY7f

Davies, Hunter. 'The Hunter Davies Interview: Dr Waldern's dream machines.' *Independent*, November 23, 1993. https://biturl.top/UjQnAb

Friedman, Vanessa. 'But Would You Wear It?' *New York Times*, June 6, 2023. https://biturl.top/rquiim

Gault, Matthew. 'Palmer Luckey Made a VR Headset That Kills the User If They Die in the Game.' *Vice*, November 7, 2022. https://biturl.top/QNbaqq

Gershgorn, Dave. 'Apple's Vision Pro AR/VR Headset: Bold, Innovative, and Ridiculously Expensive.' Wirecutter, June 6, 2023. https://biturl.top/y2IRFj

Google for Developers. 'Project Glass: Live Demo At Google I/O.' YouTube, Jun2 28, 2012. https://biturl.top/IvUbQf

Gurman, Mark. 'Apple Delays AR Glasses, Plans Cheaper Mixed-Reality Headset.' Bloomberg. January 18, 2023. https://biturl.top/Bb2U3y

Harris, Blake J. *The History of the Future: Oculus, Facebook, and the Revolution that Swept Virtual Reality*. New York: Harper Collins, 2019.

Heim, Michael. *The Metaphysics of Virtual Reality*. New York: Oxford University Press, 1993. https://biturl.top/yIfaIf

Huddleston, Tom, Jr. 'Three things Oculus co-founder Palmer Luckey splurged on when Facebook bought it for $2 billion.' CNBC, October 26, 2018. https://biturl.top/6vI7Bv

'Slower Growth for AR/VR Headset Shipments in 2023 but Strong Growth Forecast Through 2027, According to IDC.' IDC. https://biturl.top/iY7feu Accessed July 26, 2023.

Johnson, Jason. 'Atari's secret VR experiments of the 1980s.' Kill Screen. https://biturl.top/aMvENz

Kahn, Jennifer. 'The Visionary.' *New Yorker*, July 4, 2011. https://biturl.top/aeEfEz

'Oculus Rift: Step into the game.' Kickstarter. https://biturl.top/fy6rai Accessed July 29, 2023.

Lanier, Jaron. *Dawn of the New Everything: A Journey Through Virtual Reality.'* London: Penguin Random House, 2017.

Lawrence, J. M. 'Eric Howlett, 84; pioneer in virtual reality technology.' *Boston Globe*, January 15, 2012. https://biturl.top/FFVR7r

'A History of LEEP: How We Got Into VR.' LeepVR. https://biturl.top/3MVBNv Accessed July 26, 2023.

Luckey, Palmer. 'Magic Leap is a Tragic Heap.' Personal blog, August 27, 2018. https://biturl.top/Vfe6vu

Manjoo, Farhad. 'Microsoft HoloLens: A Sensational Vision of the PC's Future.' *New York Times*, January 21, 2015. https://biturl.top/uQBvqu

McCarthy, Kieren. 'Remember that amazing video of the whale leaping out the gym floor and splashing down? Yeah, it was BS.' The Register, December 9, 2016. https://biturl.top/ArQ3Qb

McFerran, Damien. 'Reality Crumbles: Whatever happened to VR?' Eurogamer, March 23, 2014. https://biturl.top/AzMJNj

MechDesignTV. 'NASA Ames – virtual environment display system, NASA's research into VR from the 1980s.' YouTube, December 4, 2018. https://biturl.top/zim2e2

Mugleston, Alex. 'Seeing Red: Looking back at Nintendo's Virtual Boy.' Gamespew. October 1, 2021. https://biturl.top/E7nYRr

Muller, Sasha. 'Google Glass Review.' Alphr, August 6, 2014. https://biturl.top/3UVzia

Nuttall, Chris. 'Magic Leap's reality check.' *Financial Times*, March 12, 2020. https://biturl.top/N7VZZf

Reckert, Clare. 'Warner signs pact to purchase Atari.' *New York Times*, September 8, 1976. https://biturl.top/67ryqa

Robertson, Adi. 'I tried Magic Leap and saw a flawed glimpse of mixed reality's amazing potential.' The Verge, August 8, 2021. https://biturl.top/6n67ju

Robertson, Adi. 'Microsoft announces new VR headsets for Windows 10, starting at $299.' The Verge, October 26, 2016. https://biturl.top/7NZR7b

Rodriguez, Salvador. 'Oculus co-founder Luckey says, "It wasn't my choice to leave" Facebook.' October 10, 2018. https://biturl.top/ZJvaum

Roettgers, Janko. 'VR Headset Owners Only Use Their Devices Six Hours a Month on Average (Study).' *Variety*, May 10, 2019. https://biturl.top/muQzMf

Roose, Kevin. 'Google Just Invested Millions of Dollars in a Very Eccentric Man.' *New York Magazine*, October 21, 2014. https://biturl.top/FRnY32

Rubin, Peter. 'The Inside Story of Oculus Rift and How Virtual Reality Became Reality.' *WIRED*, May 20, 2014. https://biturl.top/veaMFb

Russon, Charles. '"It changed the world": 50 years on, the story of Pong's Bay Area origins.' SFGate, March 10, 2023. https://biturl.top/3u2ayu

Sproull, Robert and Brock, David C. 'Oral History of Ivan Sutherland.' Computer History Museum Oral History Project. February 3, 2017. https://biturl.top/zu2aim

Suleman, Kidhr. 'Google Glass review: hands-on.' ITPro, July 18, 2014. https://biturl.top/rYfYje

Takahashi, Dean. 'Neal Stephenson & Co. turn failed Magic Leap AR project into an Audible drama.' VentureBeat, June 10, 2021. https://biturl.top/2qMNZv

Tedx Talks. 'The synthesis of imagination: Rony Abovitz and Magic Leap at TEDxSarasota.' YouTube, January 13, 2013. https://biturl.top/Ibyyye

Wingfield, Nick. 'A reality check for Microsoft's HoloLens.' *New York Times*, November 20, 2015. https://biturl.top/NvUvEj

Wiltz, Chris. 'The Story of Sega VR: Sega's Failed Virtual Reality Headset.' DesignNews, March 1, 2019. https://biturl.top/VFFVfe

Wong, Raymond. 'I used Magic Leap's AR headset, and I can tell you exactly why the company is for sale.' Inverse, March 12, 2020. https://biturl.top/7BNf2i

Cyborgs and brain computer interfaces

'Man With $6 Million "Bionic" Arm.' ABC News, June 30, 2005. https://biturl.top/UfIVza

Associated Press. 'Man's bionic arm provides hope for soldier amputees.' *Gainseville Sun*, September 16, 2006. https://biturl.top/AnqqMn

Adam, John. 'Making Hearts Beat.' Smithsonian Innovative Lives lectures, February 5, 1999. https://biturl.top/mmAnIr

'Albert Hyman, 79, Cardiologist, Dies.' *New York Times*, December 9, 1972. https://biturl.top/AjEJJr

Altman, Lawrence K. 'Arne H. W. Larsson, 86; Had First Internal Pacemaker.' *New York Times*, January 18, 2002. https://biturl.top/MzieUj

Bains, Perminder; Chatur, Safia; Ignaszewski, Maya; Ladhar, Simroop Ladhar; Bennett, Matthew T. 'John Hopps and the pacemaker: A history and detailed overview of devices, indications, and complications.' *BCMJ*, vol. 59, no. 1, January–February 2017, pp. 22–28. https://biturl.top/jYJrqm

Clynes, Manfred E. and Kline, Nathan S. 'Cyborgs and space.' *Astronautics*, September 1960. https://biturl.top/Vfqymm

Elliot, Jane. 'Bionic arm transformed my life.' BBC, July 10, 2005. https://biturl.top/FnIr2i

Fox, Margalit. 'Samuel Alderson, Crash-Test Dummy Inventor, Dies at 90.' *New York Times*, February 18, 2005. https://biturl.top/aAFnE3

Furman, Seymour. 'Early History of Cardiac Pacing and Defibrillation.' *Indian Pacing Electrophysiol J.* 2002 Jan-Mar; 2(1): 2–3. https://biturl.top/3UzMJ3

Gilbert, Daniel. 'The race to beat Elon Musk to put chips in people's brains.' *Washington Post*, March 3, 2023. https://biturl.top/raYV32

Greatbeach, Wilson. *The Making of the Pacemaker: Celebrating a Life Saving Invention*. New York: Prometheus Books, 2000.

Hanlon, Michael. 'Meet Kevin, he's the first human robot!' *Evening Herald* (Dublin), August 26, 1998.

Harbisson, Neil. *A Collection of Essays*. https://biturl.top/YNFrq2 Accessed October 14, 2023.

Harris, Francis. 'Thought powered bionic arm is a touch of genius.' *Daily Telegraph*, September 15, 2006.

Hernigou, P. 'Ambroise Paré IV: The early history of artificial limbs (from robotic to prostheses).' *Int Orthop*. 2013 Jun; 37(6): 1195–7. doi: 10.1007/s00264-013-1884-7. Epub 2013 Apr 21. PMID: 23604214; PMCID: PMC3664166. https://biturl.top/iAriuu

Hong, Jose. 'The world's first official cyborg: 10 things to know about Neil Harbisson.' *Straits Times*, May 19, 2017. https://biturl.top/beymqq

House, William. *The Struggles of a Medical Innovator: Cochlear Implants and Other Ear Surgeries*. 2011

Jabr, Ferris. 'The Man Who Controls Computers With His Mind.' *New York Times Magazine*, May 12, 2022. https://biturl.top/YBjIvu

Kobie, Nicole. 'CeBIT 2010: The computer that reads your mind.' ITPro, March 3, 2010. https://biturl.top/YZbeqi

Lee, Dave. 'Facebook team working on brain-powered technology.' BBC, April 19, 2017. https://biturl.top/mM3AF3

'Foot controlled arm.' *Life*, August 7, 1950.

Lopatto, Elizabeth. 'Elon Musk unveils Neuralink's plans for brain-reading "threads" and a robot to insert them.' The Verge, July 17, 2019. https://biturl.top/AVRvUn

Madrigal, Alexis C. 'The Man Who First Said "Cyborg," 50 Years Later.' *Atlantic*, September 30, 2010. https://biturl.top/i6BzY3

McWilliam, John A. 'Electrical Stimulation of the Heart in Man.' *Br Med J.* 1889 Feb 16; 1(1468): 348–350. doi: 10.1136/bmj.1.1468.348

Mitchell, Peter et al. 'Assessment of Safety of a Fully Implanted Endovascular Brain–Computer Interface for Severe Paralysis in 4 Patients: The Stentrode With Thought-Controlled Digital Switch (SWITCH) Study.' *JAMA Neurol.* 2023 Mar 1; 80(3): 270–278. doi: 10.1001/jamaneurol.2022.4847

Mond, Harry G.; Wickham, Geoffrey G.; Sloman, J. Graeme. 'The Australian History of Cardiac Pacing: Memories from a Bygone Era', *Heart, Lung and Circulation*, volume 21, issues 6–7, 2012 pp. 311–319, ISSN 1443–9506. https://doi.org/10.1016/j.hlc.2011.09.004

Mudry, Albert and Mills, Mara. 'The Early History of the Cochlear Implant.' *JAMA Otolaryngol Head Neck Surg.* 2013; 139(5): 446–453. doi: 10.1001/jamaoto.2013.293

Orlean, Susan. 'The man who would be a machine.' *The New Yorker,* July 26, 2022. https://biturl.top/AVfE7v

Orlowski, Andrew. '"Captain Cyborg": The wild-eyed prof behind "machines have become human" claims.' The Register, June 10, 2014. https://biturl.top/RFNZj2

Reiter, Reinhold. 'Eine neue elektrokunsthand.' *Grenzgebiete der Medizin* 1.4 (1948): 133–135.

Scott-Morgan, Peter. *Peter 2.0.*London: Penguin Random House, 2022.

University of Utah Neuroscience Initiative. 'Utah electrode array.' YouTube, February 17, 2012. https://biturl.top/3QNfm2

Warwick, Kevin. *I, Cyborg: The Cybernetic Pioneer Who Is Upgrading the Human Body – Starting with Himself.* London: Century, 2002.

Wilson Greatbatch, an oral history conducted in 2000 by Frederik Nebeker, IEEE History Center, Piscataway, NJ, USA. https://biturl.top/M3q6nm

Zeitchik, Steven. 'Why Neuralink, not Twitter, is Elon Musk's biggest challenge.' *Washington Post*, May 3, 2022. https://biturl.top/MR7vYv

Zuo, K.J. and Olson, J.L. 'The evolution of functional hand replacement: From iron prostheses to hand transplantation.' *Plast Surg (Oakv)*, 2014 Spring; 22(1): 44–51. PMID: 25152647; PMCID: PMC4128433. https://biturl.top/fieEJf

Flying cars

'Aerocar Galleries.' Aerocar for Sale. https://biturl.top/32uAna Accessed July 21, 2023.

Askari, Matthew. 'This Luxury Miami Building Boasts a Skyport That's Fit for the Jetsons.' *Architectural Digest*, August 16, 2019. https://biturl.top/aiaaQf

Barron, James. 'A 1917 Flying Boat Goes on the Auction Block.' *New York Times*, April 13, 2010. https://biturl.top/2ueiQj

Chana, William F. 'Flying Automobiles – Are They for Real?' *SAE Transactions*, vol. 105, 1996, pp. 1676–87. https://www.jstor.org/stable/44725656

'History Takes Flight.' EAA Aviation Museum, https://biturl.top/NjURn2 Accessed July 20, 2023.

Filimon, Liviu. 'Traian Vuia – the Romanian inventor who first flew a powered airplane in 1906.' *INCAS BULLETIN* 3(3): 147–150. doi: 10.13111/2066-8201.2011.3.3.15

Franklin-Wallis, Oliver. 'The battery to power Uber's flying car dreams doesn't exist (yet).' *WIRED*, August 2, 2018. https://biturl.top/YbUnqm

Gelles, David and Kitroeff, Natalie. 'Boeing and F.A.A. Faulted in Damning Report on 737 Max Certification.' *New York Times*, October 11, 2019. https://biturl.top/EvEbmm

Gilmore, Susan. 'Tired of the commute? All you need is $3.5 million.' *Seattle Times*, September 5, 2006. https://biturl.top/fEBNFb

Glenshaw, Paul. 'For a few magical years, it looked like every family would own an airplane.' *Smithsonian Magazine*, November 2013. https://biturl.top/N7RjEr

Gyger, Patrick J. *Flying Cars: The Extraordinary History of Cars Designed for Tomorrow's World*. Sparkford: Haynes, 2011.

Kasliwal, A.; Furbush, N. J.; Gawron, J. H. et al. 'Role of flying cars in sustainable mobility.' *Nat Commun* 10, 1555 (2019). https://doi.org/10.1038/s41467-019-09426-0

Martin, Chuck. 'Flying Car Gets OK to Drive on Roads.' *IOT World Today*, July 21, 2023. https://biturl.top/FVbiiq

Martin, Douglas. 'Robert E. Fulton Jr., an Intrepid Inventor, Is Dead at 95.' *New York Times*, May 11, 2004. https://biturl.top/mYny2u

'Taylor Aerocar III.' Museum of Flight. https://biturl.top/YZfuem Accessed July 21, 2023.

'Ford Flivver Airplane #1, 1926.' The Henry Ford. https://biturl.top/IfuUbq Accessed July 23, 2023.

'Waterman Whatsit.' National Air and Space Museum, https://biturl.top/EVB7vi Accessed July 23, 2023.

'Autoplane designed to run on Ground and Fly in Air.' *New Britain Herald*, 13 February 1917. https://biturl.top/rmMNJb

'Drop wreath for Brooks.' *New York Times*, March 2, 1918. https://nyti.ms/3O5V5vn

'Flying Auto Crashes.' *New York Times*, November 19, 1947. https://nyti.ms/3rH9Tty

'Glenn H. Curtiss.' *New York Times*, July 24, 1930. https://biturl.top/EvU7z2

'No sign of Brooks in beached plane.' *New York Times*, February 28, 1928. https://nyti.ms/3Y28FEC

Parks, Dennis. 'The post-war bubble.' *General Aviation News*, February 4, 2013. https://biturl.top/qQviai

Powell, Dennis E. 'Winging It! – Down the Road, Through the Clouds the Aerocar Idea Is Still Aloft.' *Seattle Times*, July 15, 1990. https://biturl.top/UnMVJn

Schäfer, A. W.; Barrett, S. R. H.; Doyme, K. et al. 'Technological, economic and environmental prospects of all-electric aircraft.' *Nat Energy* 4, 160–166 (2019). https://doi.org/10.1038/s41560-018-0294-x

Periscope Film. 'Industry on Parade.' YouTube, January 10, 2019. https://biturl.top/JVjyeu

Father of the Flying Car. Directed by Scott Hardie. (Telegraph Hill, 2002.) Amazon Prime.

'Fast-Forwarding to a Future of On-Demand Urban Air Transportation.' Uber Elevate paper, October 27, 2016. https://biturl.top/IJbmM3

Ugwueze, Ostia; Statheros, Thomas; Bromfield, Michael A.; Horri, Nadjim. 'Trends in eVTOL Aircraft Development: The Concepts, Enablers and Challenges.' AIAA 2023–2096 Session: Advanced Air Mobility and Distributed Electric Propulsion II. Jan 19, 2023. https://doi.org/10.2514/6.2023-2096

Hyperloops

Arbose, Jules. 'Britain Ends the $12-Million Hovertrain Project' *New York Times*, February 18, 1973. https://biturl.top/73IjUn

Atmore, Henry. 'Railway Interests and the Rope of Air, 1840–8.' *British Society for History of Science 2004*. https://biturl.top/vQ3mYz

Brennan, Joseph. *The Atmospheric Road: Explorations in England, Ireland, and France*. https://biturl.top/fMZzUr

'Travelling by Atmospheric Pressure.' *Brighton Gazette*, September 20, 1827.

Brunel, Isambard. *The Life of Isambard Kingdom Brunel, Civil Engineer*. London: Longmans, Green and Co, 1870. https://biturl.top/ni2eeu

Buchanan, R. A. 'The Atmospheric Railway of I. K. Brunel.' *Social Studies of Science* 22, no. 2 (1992): 231–43. http://www.jstor.org/stable/285614.

CBC News. 'High-speed rail between Calgary, Edmonton not feasible, finds committee.' May 23, 2014. https://biturl.top/q2i2mm

Clayton, Howard Francis. *The Atmospheric Railways*. Lichfield, 1966.

Connor, J. E. 'The Crystal Palace Pneumatic Tube Railway: Part 1.' *London Railway Record*, October 2003.

Connor, J. E. 'The Crystal Palace Pneumatic Tube Railway: Part 2.' *London Railway Record*, January 2004.

Daley, Robert. 'Alfred Ely Beach And His Wonderful Pneumatic Underground Railway.' *American Heritage*, June 1961. https://biturl.top/fEr22u

'Dreadful Explosion – loss of four lives.' *Freeman's Journal*, March 8, 1845.

'Travelling by Atmospheric Pressure.' *Drogheda Journal, or Meath & Louth Advertiser*, June 23, 1827.

Hadfield, Charles. *Atmospheric Railways: A Victorian Venture in Silent Speed*. Newton Abbot: David & Charles, 1967.

Hawkins, Andrew J. 'Saudi Arabia reportedly cancels deal with Virgin Hyperloop One.' The Verge, October 17, 2018. https://biturl.top/bquA7j

'The Kingstown and Dalkey Atmospheric Railway.' *Illustrated London News*, January 6, 1844.

Johnson, Eric M. 'Virgin Hyperloop hosts first human ride on new transport system.' Reuters, December 4, 2020. https://biturl.top/RBr6na

Mars, Roman (host). 'A Series of Tubes.' *99% Invisible* (podcast). September 20, 2012. https://biturl.top/JZV7fu

'The late fatal steamboat accident at Blackwall.' *Morning Herald* (London), November 16, 1844.

Murray, Kevin. 'The Atmospheric Railway to Dalkey.' *Dublin Historical Record*, vol. 5, no. 3, 1943, pp. 108–20. https://www.jstor.org/stable/30080115

Musk, Elon. 'Hyperloop Alpha.' Tesla website. https://biturl.top/Vbiiqq

'Obituary record: Alfred Ely Beach.' *New York Times*, January 2, 1896. https://nyti.ms/3rQncrp

'Underground Tube Is Urged for High-Speed Trains in Northeast: Pneumatic Pressure and Gravity Would Propel the Cars', *New York Times*, August 1, 1965.

'HyperloopTT signs deal for first test track in China.' Reuters, July 19, 2018. https://biturl.top/Vn2Eri

Salter, Robert M. *The Very High Speed Transit System*. Santa Monica, CA: RAND Corporation, 1972. https://biturl.top/zqe6ry

'The pneumatic tunnel under Broadway, New York.' *Scientific American*, February 19, 1870. https://biturl.top/AVVJny

'An atmospheric railway.' *St. Pancras Guardian and Camden and Kentish Towns Reporter*, January 28, 1910.

Temperton, James. 'The strange tale of the hovertrain, the British hyperloop of the 1970s.' *WIRED*, July 2, 2018. https://biturl.top/Qn2iqq

Thompson, Clive. 'The Hyperloop Will Be Only the Latest Innovation That's Pretty Much a Series of Tubes.' *Smithsonian Magazine*, July 2015. https://biturl.top/2aiMBr

Valdez, Jonah. 'Elon Musk's Hyperloop prototype tube is gone. What does it mean for his tunneling dream?' *Los Angeles Times*, November 5, 2022. https://biturl.top/miuuMf

Vance, Ashlee. *Elon Musk: How the Billionaire CEO of SpaceX and Tesla is Shaping Our Future*. London: Ebury Publishing, 2015.

Wade, John. *Transport Curiosities: 1850–1950*. Philadelphia: Pen and Sword Books, 2022.

Wickens, A. H. 'APT with hindsight.' Personal site. https://biturl.top/i2mMJv

'2016 Local Content Vehicles Cities Tour: Pueblo Railway Museum.' C-SPAN, September 31, 2016. https://biturl.top/6BZfaa

Smart Cities

Booth, Robert. 'Oil rich Gulf emirate plans city powered by renewable energy.' *The Times* (syndicated in *Calgary Herald*), May 13, 2007.

Brown, Jennings. 'Privacy Expert Resigns From Alphabet-Backed Smart City Project Over Surveillance Concerns.' Gizmodo, October 23, 2018. https://biturl.top/uU3uem

Brown, Will. 'The origins of the smart city.' Medium, January 15, 2021. https://biturl.top/q2UzMb

Canadian Press. 'The conscientious computer.' *Northern Sentinel*, 30 April, 1969.

Canadian Press. 'Sidewalk Labs' advisory panel member resigns citing "profound concern" about project.' CBC News, October 5, 2018. https://biturl.top/FziUFb

Casciato, Leonard and Cass, Sam. 'Pilot Study of the Automatic Control of Traffic Signals by a General Purpose Electronic Computer.' 1962. https://biturl.top/rmuqi2

Canon, Gabrielle. '"City of surveillance": privacy expert quits Toronto's smart-city project.' *Guardian*, October 23, 2018. https://biturl.top/2e63Uj

'Cisco and City of San Francisco Embark on "Sustainable 21st Century San Francisco" Project.' Cisco, press release, December 1, 2009. https://biturl.top/EB3yY3

'Cisco and the City of Seoul Launch Personal Travel Assistant Solution.' Cisco, press release, May 21, 2009. https://biturl.top/NNRzue

'Cisco's Connected Urban Development Program Signposts the Future Era of Sustainable Work.' Cisco, press release, September 23, 2008. https://biturl.top/BzeAvi

Collins, George R. 'The Ciudad Lineal of Madrid.' *Journal of the Society of Architectural Historians*, vol. 18, no. 2 (May, 1959), pp. 38–53. https://doi.org/10.2307/987976

Cugurullo, Federico. *Megaurbanisation in the Global South*. Taylor & Francis, 2016.

deMause, Neil. 'How The 1977 Blackout Was Bushwick's Grimmest Moment.' Gothamist, September 28, 2016. https://biturl.top/7jiaqy

Descalsota, Marielle. 'Malaysia's $100 billion luxury estate was supposed to be a "living paradise." Instead, 6 years into development, it's a ghost town full of empty skyscrapers and deserted roads – take a look.' Insider, June 8, 2022. https://biturl.top/eQJzY3

Dineen, J. K. 'As S.F.'s housing production stalls, this development continues to add affordable homes.' *San Francisco Chronicle*, May 29, 2023. https://biturl.top/ERrQzq

Dirks, Suzanne and Keeling, Mary. 'A vision of smarter cities.' IBM Institute for Business Value, December, 2009. https://biturl.top/nEzmAj

Doctoroff, Daniel L. 'My next chapter: Fighting ALS.' Medium (Sidewalk Talk), December 16, 2021. https://biturl.top/J7JZR3

Doctoroff, Daniel L. 'Why we're no longer pursuing the Quayside project – and what's next for Sidewalk Labs.' Medium (Sidewalk Talk), May 7, 2020. https://biturl.top/FVNrue

Dworkin, Larry. 'Toronto's computer traffic cop eases stop-light snarls.' *Star-Phoenix*, 14 September, 1967.

Ferrer, Josep-Ramon. 'Barcelona's Smart City vision: an opportunity for transformation.' *Field Actions Science Reports*, Special Issue 16, 2017, 70–5. https://biturl.top/Yziaee

Flood, Joe. *The Fires: How a Computer Formula Burned Down New York City – and Determined the Future of American Cities*. London: Penguin Books, 2010.

'About Forest City.' Forest City. https://biturl.top/ryyUFb Accessed October 18, 2023.

Forrester, Jay W. *Urban Dynamics*. Cambridge, Massachusetts, MIT Press, 1969.

'Projects: Masdar City.' Foster + Partners. https://biturl.top/2aqAr2 Accessed August 23, 2023.

Frey, Christopher. 'World Cup 2014: inside Rio's Bond-villain mission control.' *Guardian*, May 23, 2014. https://biturl.top/7JneIb

Gaffney, Christopher and Robertson, Cerianne. 'Smarter than smart: Rio de Janeiro's Flawed Emergence as a Smart City.' *Journal of Urban Technology*, volume 25, 2018: pp. 47–64. https://doi.org/10.1080/10630732.2015.1102423

Goldenberg, Suzanne. 'Masdar's zero-carbon dream could become world's first green ghost town.' *Guardian*, February 16, 2016. https://biturl.top/Q7ZRj2

Greenfield, Pamela Licalzi. 'Korea's High-Tech Utopia, Where Everything Is Observed.' *New York Times*, October 5, 2005. https://biturl.top/ZVbQZn

Günel, Gökçe. *Spaceship in the Desert: Energy, Climate Change, and Urban Design in Abu Dhabi*. London: Duke University Press, 2019.

Hian, Kerr Puay. 'Woman from China buys JB Forest City condo unit with "unparalleled sea view", only to find a "ghost town".' Mothership, February 17, 2023. https://biturl.top/RneyMr

'How the Irish changed the traffic laws in Tipperary Hill, Syracuse.' IrishCentral, May 25, 2023. https://biturl.top/3u6Fbi

James, Ian. 'Songdo: No Man's City.' *Korea Expose*, October 15, 2016. https://biturl.top/MJFzee

Kahn, Rahma. 'Is this the city of the future?' *Independent*, 16 May 2021.

Kobie, Nicole. 'How London became the first smart city back in 1854.' CityMonitor, September 19, 2016. https://biturl.top/vErmIb

Kobie, Nicole. 'Ricky Burdett: living in an urban age.' Deutsche Bank Insights, May 2, 2019. https://biturl.top/Zz2Mfe

Kobie, Nicole. 'Malaysia's city of the future is an uncanny valley.' *WIRED*, March 22, 2016. https://biturl.top/a226ji

Los Angeles (Calif.) Community Analysis Bureau. 'The state of the city report, a cluster analysis of Los Angeles.' June 1974.

Malapert, Etienne. 'The City of Possibilities.' https://biturl.top/r6Vv2y Accessed August 23, 2023.

Mattern, Shannon. *A City is Not a Computer: Other Urban Intelligences*. Princeton: Princeton University Press. 2021.

McShane, Clay. 'The Origins and Globalisation of Traffic Control Signals.' *Journal of Urban History*, vol. 25 no. 3, March 1999 379–404. https://biturl.top/2M7JB3

'Traffic decongestant gets the green light.' *National Post*, March 8, 1980.

'7 Children Dead in Brooklyn Fire.' *New York Times*, March 7, 1970. https://biturl.top/JjYNfy

O'Connell, Mark. *To Be a Machine: Adventures Among Cyborgs, Utopians, Hackers and the Futurists Solving the Modest Problem of Death*. London: Granta, 2017.

O'Kane, Josh. *Sideways: The City Google Couldn't Buy*. Toronto, Canada, Random House Canada, 2022.

'Signal computer aids Toronto's traffic tie-ups.' *Ottawa Citizen*, 19 August, 1964.

Pearson, Jordan. 'Sidewalk Labs' 1,500-Page Plan for Toronto Is a Democracy Grenade.' *Vice*, June 24, 2019. https://biturl.top/rYrAbq

Pettit, Harry and White, Chris. 'A glimpse into the future? $39 billion high-tech smart city in South Korea turns into a "Chernobyl-like ghost town" after investment dries up.' *Daily Mail*, March 28, 2018. https://biturl.top/7nuaqq

Poon, Linda. 'Sleepy in Songdo, Korea's Smartest City.' Bloomberg, June 22, 2008. https://biturl.top/i2IbYf

Revkin, Andrew Revkin. 'Solar City in Gulf Could Set Standard.' *New York Times* service in *Miami Herald*, February 16, 2008.

Robbins, Jim. 'Why the Luster on Once-Vaunted "Smart Cities" Is Fading.' *Yale Environment* 360, December 1, 2021. https://biturl.top/3IBJzm

Salmon, Andrew. 'Breaking ground on a Korean bid to rival Shanghai.' *New York Times*, November 12, 2004. https://biturl.top/aaUjei

Shubber, Kadhim. 'Tracking devices hidden in London's recycling bins are stalking your smartphone.' *WIRED*, August 9 2013. https://biturl.top/IfaaYz

Singer, Natasha. 'Mission Control, Built for Cities.' *New York Times*, March 3, 2012. https://biturl.top/F7FBry

Swabey, Pete. 'IBM, Cisco and the business of smart cities.' *Information Age*, February 23, 2012. https://biturl.top/jyY3Un

Townsend, Anthony M. *Smart Cities: Big Data, Civic Hackers and the Quest for a New Utopia*. New York: Norton, 2014.

Turner, Chris. *How to Breathe Underwater: Field Reports from an Age of Radical Change*. Windsor: Biblioasis, 2014.

Vallianatos, Mark. 'How LA Used Big Data to Build a Smart City in the 1970s.' June 22, 2015. https://biturl.top/q2Ib6v

Villa, Nicola and Wagener, Wolfgang. 'Connecting Cities: Achieving Sustainability Through Innovation.' Cisco Connected Urban Development Global Conference 2008. https://biturl.top/j6Rvuu

Vives, Antoni. *Smart City Barcelona: The Catalan Quest to Improve Future Urban Living*. Eastbourne: Sussex Academic Press, 2018.

Wainwright, Oliver. 'Nine million people in a city 170km long; will the world ever be ready for a linear metropolis?' *Guardian*, September 8, 2022. https://biturl.top/U7byE3

Walsh, Bryan. 'Masdar City: The World's Greenest City?' *Time*, January 25, 2011. https://biturl.top/bMvQJv

Outro

Kobie, Nicole. 'Monzo has a cunning plan to stop you wasting money on junk food.' *WIRED*, September 6, 2019. https://biturl.top/nueuQ3

Index

1X Technologies - EVE and NEO 142
510 Systems 45–6

A* algorithm 124
Abovitz, Rony 174–5
Ackland, Nigel 197
activism 311, 331
ADALINE and MADALINE neural networks 88
Advanced Micro Devices (AMD) 17
Advanced Research Projects Agency Network (ARPANET) 15, 16
Al Jaber, Sultan 304
Aldebaran Robotics 138–40
Alderson, Samuel W. 194–5
Alef Aeronautics 244–5
Allende, Salvador 300
Alphabet 49, 63, 65
  Sidewalk Labs 307–13, 331
  see also Google
Alphago algorithm, DeepMind 94
Altman, Sam 65
Amazon 67, 91, 96
American Institute Fair (1867) 263–4
American Research and Development Corporation (ARDC) 16
American Society of Automotive Engineers (SAE) 20
Anderson, Sterling 49
Anthony's Robots 45–6
Antonov, Michael 168
Apple 15, 16, 17, 67, 96, 104, 159, 171, 324–5
  augmented reality headsets 176–8, 180
APT-E (Advanced Passenger Train) project, British Rail 269–70
Arcadio (aka Stelarc), Stelios 210
Argo AI 49
ARPA *see* DARPA (Defense Advanced Research Projects Agency)
ARPANET (Advanced Research Projects Agency Network) 15, 16
artificial intelligence (AI) 10, 11, 12, 15, 32, 34, 315
  Alan Turing 72–4
  Alphabet/Google 65, 78, 92, 93, 98, 99–100, 104–5, 106
  'Blocks World' 84
  chess challenges 78, 81, 84, 94
  computer vision 69–70
  concerns and misuse 65–6, 68–9, 70, 95–8, 99–100, 102–8, 212, 327–8
  data resources and ImageNet 91–2
  deep and machine learning 69, 85, 89, 90–1, 92–7
  'expert systems' and rules based models 67, 81–3, 329
  'general' intelligence 12, 68–9, 70–1, 82–3, 100–2
  graphics processing units (GPUs) 93
  John McCarthy 66–7, 75–7, 78, 79, 80, 83, 126
  large language models (LLM) 98–106
  Lighthill report (1973) and opposition 78–81, 84
  narrow AI 69, 70, 81
  natural language processing (NLP) 69–70, 91
  neural networks 67, 69, 84–90, 94
    convolutional 92
  OpenAI and ChatGPT 65, 67–8, 83, 98, 100–2, 105, 142
  Perceptron model 86–8
  pre-1970s 66–8, 72–8
  SRI Shakey robot 15, 119–26
  strong vs weak 68, 70
  symbolic 69, 88
  Timnit Gebru *et al* AI paper controversy 104–7, 331
Asimo robot, Honda 136–7
ASKA 241
Asseily, Alex 244
Atari 158–9, 160
augmented reality (AR) and virtual reality (VR) 9, 11, 144
  Apple 176–8, 180, 325
  Atari 158–9, 160
  computer graphics quality 155–6, 157
  development costs 180
  Eric Howlett and the LEEP system 157–8, 161–2
  Facebook and Meta 168, 170, 176, 180
  Google Glass smart spectacles 147–8, 149, 172

Ivan Sutherland VR headsets 149–54, 332–3
Jaron Lanier 159, 160–1, 165
Jonathan Waldern 163–4
metaverse the 178–80
Microsoft HoloLens 176
motion sickness 165, 166, 167, 168, 169–70, 171, 176
NASA 157, 158, 159, 161–3
Nintendo and Virtual Boy 165–6
Palmer Luckey and Oculus Rift headsets 166–8, 169, 170, 175
PTSD treatment 172
Rony Abovitz and Magic Leap 174–5
Samsung Gear VR 168, 170
Sebastian Thrun 172–3
Sega VR 164–6, 180
Snap and Vergence Labs 176
Sony PlayStation VR 169–70
Virtual Visual Environment Display (VIVED) 161–3
VPL DataGlove and EyePhone 161, 162, 163, 165
Aurora 49, 57
Aviauto roadable aircraft 229

Babbage, Charles 13
Bagnell, Drew 49
Baidu 92
Bakken, Earl 190, 191
Balsillie, Jim 311
Banavar, Guru 298
banking industry 326–7, 328–9
Barcelona, Spain 317–18
Bard LLM, Google 98, 99
BASIC (Beginners' All-purpose Symbolic Instruction Code) 15
battery technology 13, 193, 239–40
Beach, Alfred 263–4, 326
Becker, Allen 165–6
Behringer, Reinhold 19, 33, 34–5, 36–7, 38, 39, 40
Bel Geddes, Norman 10, 21–4, 226, 332
Bell
Helicopter 151–2
Labs 14, 16
Nexus 240, 241
Bender, Emily 104–5
Bengio, Yoshua 92
Benz, Carl 220
Berlichingen, Götz von 194
Berners-Lee, Tim 15
Bertin, Jean 270–1
Bezos, Jeff 330
bionic technology 195–6
Blackrock Neurotech 211–12
'Blocks World' AI development 84
Blomfield, Tom 326
'Blue Cruise' automation system, Ford 11
Boeing 240, 241, 246
Bombe and Colossus decrypting machines 14, 74
Boring Company 22, 277
Boston Dynamics 12, 16, 109, 111–12, 127–34, 137, 143, 144
Atlas robot 131, 132, 143
BigDog robot 130, 132
RHex 130
Sandflea 130
Spot 132, 133, 137, 144
Stretch 132, 143
Bostrom, Nick 70–1
BrainGate/Utah Array implants 200–2, 210–11
Branson, Richard 278
Bria, Francesca 318
Brin, Sergey 147–8
British army 2222
British Medical Journal 189
British Rail 266, 269–70
Brooks, Harry 222–3
Brooks, Rodney 83
Brunel, Isambard Kingdom 259
Bryan, Leland 230
Buchanan, Bruce 82
Buolamwini, Joy 104, 105, 327–8
Burdett, Ricky 316–17

Canella, Judge John 295
carbon emissions 105–6, 238, 239, 304
Cardboard, Google 170
Carmack, John 167
Carnegie Mellon University (CMU) 17, 39, 42–3, 44–5, 77, 84
Carson, Johnny 116–17
Catalyst Research Corporation 193
Catanzaro, Bryan 93
CAVForth project, Forth Bridge 52–3
Cavoukian, Ann 311–12
CeBIT (2010) trade show 181–2
Central Intelligence Agency (CIA) 153–4
Cereproc 208
Chambless, Edgar 316
Chan Wolf, Helen 120, 125
ChatGPT 67–9, 99, 100–1, 105
Cisco 296–7, 302
Claydon, Tony 268–9
Clegg, Samuel 255, 256
climate crisis 105–6, 330
Clinton Foundation 296–7

Clynes, Manfred 184
cochlear implants 185–8
Commodore 15
Convair (Consolidated Vultee Aircraft) 225–7, 230
convolutional neural networks 92
Cook, Tim 177
Cornelius, Nancy 129
COVID-19 global pandemic 49, 179, 310, 331
Cray 1 supercomputer 15
Crevier, Daniel 66, 81, 86
Crow, Steven 233
Croydon Times 114
Cruise, GM 49, 58–9, 61, 63, 331
Cubitt, Sir William 259
Cummings, Bob 229
Curtiss, Glenn 221
Cyberjaya, Malaysia 313–14
cybernetics 74–5
cyborgs and brain computer interfaces (BCIs) 11
  BCIs 210–16
  Blackrock Neurotech 211–12
  Elon Musk and Neuralink 182, 183–4, 210, 212–13, 216
  g.tec intendiX interface 182, 214–15
  Kevin Warwick 198–204
  Mark Zuckerberg and Facebook 182, 210, 213–14, 216
  medical applications 182–3, 185–98, 204, 207–9, 211, 212, 214, 215–16
  Neil Harbisson and colourblindness 204–7
  origins of the term 'cyborg' 184
  pacemakers 189–93
  Peter Scott-Morgan and ALS treatment 207–9
  prosthetics and bionic technology 193–8
  RFID chips 183, 199–200, 204
  Utah Array/BrainGate implant 200–4, 210–11
  William House and cochlear implants 185–8
Cyc AGI 82–3

da Vinci, Leonardo 110–11, 220
Daimler-Benz/Mercedes-Benz 34–5, 38
Dally, Bill 93
DARPA (Defense Advanced Research Projects Agency) 16, 19, 31, 60, 77, 83, 119, 130–1, 132, 196, 198, 325, 332
  Grand Challenge 40–2, 60
DataGloves, VPL Research 161, 162
Davis, Ruth 120
Dean, Jeffrey 104
DEC 81–2
Deep Blue AI 81, 84
DeepMind, Google 83, 92, 94
DeGray, Dennis 211
Dendral algorithm 82
Desforges, Abbé 220
Devol, George 115
Dickmanns, Ernst 19, 32–9, 40, 42, 323
Dietrich, Carl 233–4
Difference Engine 13
Digital Equipment Corporation (DEC) 16
Dirks, Susanne 297–8
diversity, lack of tech industry 326–8
Djourno, Dr André 185
DNNresearch 92, 94
Doctoroff, Dan 308, 309, 311
Dolgov, Dmitri 45
Dreyfuss, Henry 226
driverless cars/technology 10, 19–20, 124, 329–30
  4D system 36–8
  Anthony Levandowski 45, 46, 48–9
  Anthony's Robots pizza delivery 45–6
  Argo AI 49
  artificial intelligence (AI) 32, 34, 70
  autonomy spectrum 20, 57–8
  current motivation for innovation 62–3
  DARPA and DARPA Grand Challenge 40–2, 60
  Ernst Dickmanns and VaMoRs 32–9, 42, 323
  fatalities 47, 53–6, 59, 327
  General Motors and Cruise (RCA) 24–6, 49, 58–9, 331
  global cost of development 31
  Google/Alphabet and Waymo 45, 46–8, 49, 50–1, 54, 58, 59, 60, 324, 331
  motorways 21–2, 23–4, 26–7, 29–30, 35
  NavLab 5 System and RALPH 39–40
  'no hands across America' trip 39
  Norman Bel Geddes 22–4
  Prometheus project 33–5
  public transport 51–3, 61, 63–4, 305–7
  Road Research Laboratory (RRL), UK 26–30
  Tesla 30–1, 47, 49, 61
  TRL GATEway pods 51–2
  Tsukuba Mechanical Engineering Lab 39
  Uber 48, 49, 53–7

US Congress 39
VaMP and VITA-2 35
World's Fair, New York (1939) 20–1
Dugan, Regina 214

Electronic Age magazine 24–5
Electronic Numerical Integrator and Computer (ENIAC) 14, 73
Eliot, George 262
ELIZA 78
Elmqvist, Rune 190
Em, David 157
EMIEW, Hitachi 137
Engelberger, Joe 115, 117
Engineer magazine 262
Enigma cypher 74
Entscheidungsproblem 72–3
ERCO Ercoupe 224–5
Ermisse, Daniel 270–1
Evans, Dave 154
Exhibition of the Society of Model Engineers (1928) 112–13
Eyriès, Dr Charles 185–6

Facebook 98, 168, 170, 213–14, 216
see also Meta
Fairchild Semiconductor 14, 16–17
Federal Aviation Administration (FAA), US 217, 227, 233, 240, 246
Fei-Fei Li 91, 104
Feigenbaum, Edward 82
Fenton, Robert 30, 31
Ferrer, Josep-Ramon 317–18
Fifth Generation Computer Project, Japan 83
Firefly driverless car 47, 48, 63
Fisher, Scott 159, 161
Flood, Joe 293–5
Flow, Sidewalk Labs 312
flying cars 9, 11, 77, 217–19, 329
air-traffic control 242–3
Alef Aeronautics Model A 244–5
battery technology 239–40
business model operation 243–4
Carl Dietrich – Terrafugia Transition 233–4
EHang 184 (AAV) 237
Glenn Curtiss – Autoplane (1917) 221–2
Hafner Rotabuggy 222
Henry Ford – Flivver 222–3, 230
Henry Smolinski Cessna/Ford 230
Jaunt Air Mobility Journey 237
Juan de la Cierva – autogyro (1923) 222
Ken Wernicke – AirCar 233
landing space and vertiports 241–2
Lilium Jets 217, 234–6, 237, 239, 241–2, 244, 246–7
Luigi Pellarini – Aeronova 230
Moulton Taylor – Aerocar 228–9, 230
Opener BlackFly (later Pivotal Helix) 245
Paul Moller inventions 231–3
post-WWII small aircraft sales 224–5
potential practical applications 247–8
regulatory authorities 217, 227, 233, 240–1, 246
René Tampier – Avion-automobile (1921) 222
'roadable aircraft' 225–30, 244–5
Robert Edison Fulton – Airphibian 227–8
Sebastian Thrun – Kitty Hawk 245–6
Steven Crow – Starcar 4 233
Theodore Hall and the ConvAirCar 225–7, 230
Trajan Vuia – aeroplane-automobile (1902) 220–1
Vladimir Tatrinov – Aeromobile (1909) 221
Volocity – Volocopter 236–7
VTOLs and eVTOLs 219, 232–3, 235–48
Waldo Waterman inventions 223–4
Forbes Nash Jr, John 77
Ford, Henry 220, 222–3, 230
Ford Motor Company 11, 222–3, 229
Forest City, Malaysia 283
Forrester, Jay 290
Fortnite 179
Fortran 15
Foster + Partners 304, 305, 314
Foster, Quintin 149–51, 153
Fox Dunn, Angela 124
Fujitsu 137
Fulton, Robert Edison 227
'Futurama' exhibition, World's Fair (1939) 21–2

Gaffney, Christopher 299
gaming industry 92–3, 96, 124, 156–7, 163–70, 197
see also augmented reality (AR) and virtual reality (VR)
Garrett AiResearch 272
Gasson, Mark 202–3
Gates, Bill 125
Gebru, Timnit 104, 327–8, 331
General Motors (GM) 24, 49, 111, 115–16, 118, 125, 331
General Problem Solver 78

generative adversarial network (GAN) 94
Gerhardt, Marcus 211
Glass smart spectacles, Google 147–8, 172–3
Goddard, Robert 265
Goodfellow, Ian 94
Google 45, 159, 312
  artificial intelligence (AI) 78, 92, 94, 99–100, 104–5, 106, 331
  augmented reality and Glass smart spectacles 147–8, 170, 172–3
  driverless cars 46–9, 50–1, 54, 58, 59, 60, 324, 331
  robots/robotics 127, 131, 132
  X Labs 323–4, 330
Gow, David 195
Graphics Processing Units (GPUs) 13, 93, 156
Greatbatch, Wilson 191–2
Greenfield, Adam 319–20
Gregory, Richard 79
Grumman TACRV (Tracked Air Cushion Research Vehicle) 272
g.tec 182, 214–15
the Guardian 298, 316
Guger, Dr Christoph 214–15
Günel, Gökçe 305, 306–7

Hall, Dave and Bruce 41, 44, 47
Hall, Theodore 225–7, 230
Harbisson, Neil 204–7
Hart, Peter 119, 121, 123, 124, 125
Harvard University 149, 150–4
Hashme, Shariq 144
Hassabis, Demis 94
Hebb, Donald 86
Heilig, Morton 151
Heim, Michael 155–6
Henn na Hotel, Japan 144–5
Herzberg, Elaine 53–6, 327
Heseltine, Michael 268
Hilbert, David 73
Hinton, Geoffrey 89, 90–1, 92, 99–100
Hitachi 137
Hoff, Marcian 88
Hommet, Christophe 234, 2241–242
Honda 111, 134, 143
Hopps, John 189–90
House, William 185–8
Howlett, Eric 157–8
HTC Vive VR headset 169
Hyman, Albert 189
hyperloops 11, 280–2, 332
  20th century pneumatic trains 265–71
  Beach Pneumatic Transit Company 263–4
  cable systems 254
  Dalkey Atmospheric Railway (DAR) 257–9
  Elon Musk 249, 250, 251, 274–6
  Eric Laithwaite and Tracked Hovercraft 265–9, 270
  European Hyperloop Week (EHW) (2023) 249–50, 274
  George Medhurst and Victorian pneumatic transport 250–1, 252, 253, 254
  Hyperloop TT 278
  Isambard Kingdom Brunel 259
  Jean Bertin and Aérotrain 270–1
  John Vallance's air-propelled carriage, Brighton (1826) 253–4
  maglev technology 272–3
  pneumatic postal shuttles 252–3, 260–1
  Rammel and Latimer Clark 252, 260–3
  Rohr Industries 271, 272
  Samuda brothers and Samuel Clegg 255–9
  Shinkansen 266, 280, 281
  Tracked Air Cushion Vehicles (TACV) programme 271–2
  TransPod 278–80
  Victorian atmospheric and pneumatic transport 250–3, 254, 255–64
  Virgin Hyperloop One 278

IBM 84, 125, 194–5, 297–8
ImageNet 91–2
Industry on Parade TV programme 228
the Information website 175
Ingels, Bjarke 314
Intel 16, 93
International Conference on Robotics and Automation (2023), IEEE 109–10, 128
internet 9–10
Iribe, Brendan 167–8
iRobot 84
Ishiguro, Hiroshi 109, 140–1

Jaguar i-Pace 50–1
Joby Aviation 233, 237

Kalinske, Tom 166
Kasparov, Garry 84
Kates, Josef 289
Kawasaki Heavy Industries 117–18
Kay, Alan 154, 159
Keeling, Mary 297–8
Kese, Peter 205

Kim, John 301–2
Kitano, Hiroaki 142
Kitty Hawk 245–6
Kline, Nathan 184
Kobrinski, Alexander 195
Koppen, Otto 222
Krizhevsky, Alex 92
Krueger, Myron 155
Kuala Lumpur, Malaysia 314
Kuiken, Dr Todd 195–6

La Cierva, Juan 222
Laithwaite, Eric 266–7, 272
languages, early computer 15
Lanier, Jaron 159–61, 165
Larsson, Arne 190
Latimer Clark, Josiah 252, 260
Laurel, Brenda 159
Laverde, Alberto Vejarano 190
Le Corbusier 316
LeCun, Yann 92
Leg Lab 127–9
Legg, Shane 94
Lenat, Doug 82–3
Levandowski, Anthony 40, 42, 45, 46, 48–9, 63
Lidwill, Mark 189
Life magazine 65, 125
Lighthill, James 78–81, 84
Lilium 217, 234–6, 237, 239, 241–2, 244, 246–7
Linden Lab 166
the Line, Saudi Arabia 315
linear cities 315–16
linear induction motor (LIM) 267, 271, 272
Linear Induction Motor Research Vehicle (LIMRV) 272
Link, Sidewalk Labs 312
lithium batteries 13, 240
Llama, Facebook 98
Lobban, Joan 268
Lockheed Missiles 265
London Pneumatic Despatch Railway (LPDR) 260
longtermism 70–1
Los Angeles Community Analysis Bureau (CAB) 291–2
Lovelace, Ada 13
LS3, Boston Dynamics' 130
Luckey, Palmer 166–8, 169, 170, 175
Lyft 60
Lyons 14

M-PESA 328–9
MacWilliam, John Alexander 189
Macy's 224–5
Magic Leap 174–5
maglev technology 272–3
Mahan, Steve 48
Maisonnier, Bruno 138–40
Malapert, Etienne 304
Manchester Mark 1 74
Markoff, John 46
Marsh, Burton 288
Masdar City, UAE 303–7
Maslow, Abraham 329
Mattel 166–7
Mattern, Shannon 285, 320
Mauchly, John 73
McCarthy, John 66–7, 75–7, 78, 79, 80, 83, 126
McClelland, James L. 89
McCulloch, Warren 77, 85–6
McGreevy, Michael 161
McMillan-Major, Angelina 105
McShane, Clay 288
Mechanical Turk, Amazon 91
Mechanics Magazine 262
Medhurst, George 251, 252
Medtronic 191, 192
Meta 79, 176
metaverse 178–80
Michie, Donald 81, 84
Microsoft 104
  artificial intelligence (AI) 92, 99, 100–2
  HoloLens 176
  Windows 15
military, US 16, 87, 112, 119–20, 127, 129, 130–1, 132, 150, 153–4, 162, 196, 198, 222, 228, 294, 325, 330–1
Minadoi, Giovanni Tommaso 194
Minsky, Marvin 65, 66–7, 76, 77, 88, 160
MIT (Massachusetts Institute of Technology) 77–8, 127–8, 157, 159, 198
Mitchell, Claudia 196
Mitchell, Margaret 104–5, 331
Mitchell, Nate 168
Mitsubishi 137
Moller, Paul 231–3
Montandon, Adam 205
Montemerlo, Mike 45
Monzo 326–7
Moore and Moore's Law, Gordon 93
Morgan, Garret 287
Musk, Elon 22, 65, 79, 98, 111, 141, 182, 183–4, 210, 212–13, 249, 250, 251, 325, 330–1
Mustang Mach-E, Ford 11
MYCIN 82, 83

Nagle, Matt 211
Naimark, Michael 159, 170
NASA (National Aeronautics and Space Administration) 157, 158, 159, 161–3, 325
National Transport Safety Board (NTSB), US 53
Nature 238
NerveGear headset 171
NetJets 244
neural networks 9, 13, 34–5, 84–90, 94
  convolutional 92
Neuralink 182, 212–13, 216
New Scientist 99
New York City fires, 1970s 293–4
New York Times 46, 87, 170, 176, 224, 227, 264, 266, 298
Newell, Allen 77, 78
Ng, Andrew 93
Nilsson, Nils 122
Nintendo 171
  Virtual Boy 165–6
Normann, Richard 200, 211
Nvidia 93

Ocado Technologies 143
Oculus Rift AR smart glasses 149
Oculus Rift VR headsets 166–9, 170
O'Hagan, John 294, 295
O'Kane, Josh 308, 309
Okhitovich, Mikhail 316
Olson, Karl 30, 31
Open Bionics 197–8
OpenAI 65, 67–9, 83, 98–9, 100–2, 142
Opener BlackFly/Pivotal Helix 245
Optimize3D 208
Optimus robot, Tesla 141
Osborne 1 15
Otto 48
Ottobock 197

pacemakers 189–93
Paes, Eduardo 298–9
Page, Larry 132, 172, 237, 245, 307–10, 325–6
Papert, Seymour 88
Paré, Ambroise 194
PDP-8 and PDP-10 'microcomputers' 15
Pellarini, Luigi 230
Penoyre, Slade 29
perceptrons, AI 88
Permobil F5 Corpus 209
Petman, Boston Dynamics' 131
Pichai, Sundar 65, 175
Pinkus, Henry 255
Pitcairn, Harold 222
Pittau, Rev. Joseph 135
Pitts, Walter 85–6
Playter, Robert 129
Pliny the Elder 193–4
Pombo, Jorge Reynolds 190
Popular Mechanics magazine 226
Popular Science magazine 164
Post Office 252, 261
Pravdich-Neminsky, Vladimir 210
Presper Eckert, J. 73
Princeton Press-Courier 26
ProPublica software 96
prosthetics and bionic technology 193–8
Prototype This! TV programme 45–6

Qualcomm 31
Quigg, Doc 26

radio 12, 23, 113
Radio Corporation America (RCA) 24–6
radio frequency identification (RFID) 12
Raibert, Marc 109, 127–34
RAND (Office of Scientific Research and Development) 16
RAND Corporation 294–6
Rander, Peter 49
Ratio Club 86
Reflection Technologies Inc (RTI) 165–6
the *Register* 41, 202
Reichardt, Jasia 109
Reiter, Reinhold 194
RFID chips 183, 199–200, 204
Richards, William 112
Rio de Janeiro, Brazil 298–300
Rising Sun film (1993) 129
Rizzo, Skip 172
Roach, Ron 272
Road Research Laboratory (RRL), UK 26–9, 60
robots/robotics 12, 15, 16, 41, 80–1, 84
  Agility Robotics' Digit 142
  Aldebaran Robotics - Nao and Pepper 138–40
  Boston Dynamics and Leg Lab 12, 16, 109, 111–12, 127–34, 137, 143, 144
  carer roles 132, 137, 138, 145
  early 112–15
  Elektro – Sparko the dog 113–14
  Engineered Art – RoboThespian 112, 140
  Hanson Robotics - Sophia 112, 140
  Hiroshi Isiguro 140–1

Honda – Asimo and Avatar Robot 112, 134–7, 141
humanoid and bipedal 110–11, 112, 129, 130–1, 134–44, 145–6, 332
ICRA (2023) 109–10, 128
Japanese 112, 134–8, 141–2, 144–5
jobs/industrial uses 110, 115–18, 131–2, 138, 139–40, 142–5
Knightscope K5 – Steve the robo-guard 112
military use and law enforcement 119–20, 127, 129, 130–1, 132
Prosper Robotics – Alfie 144
Robear carer robot 145
Sony AIBO robotic dog 137
SRI – Shakey robot 118–26, 325
Sunnyvale - Figure 01 robot 142
surgical uses 110, 129
Tesla – Optimus 141
Toyota – Robina and T-HR3 robots 138
UBTech – Walker robot 142
Unimation – Unimate One and PUMA 115–18
William Richards and 'Eric' 112–14
Rochester, Nathaniel 76
Rock, Arthur 16
Rohr Industries 271–2
Roomba 84
Rosedale, Philip 166
Rosen, Charlie 120, 123, 124–6
Rosenblatt, Frank 86–8
Rumelhart, David E. 89
Russell, Ben 114

Salesky, Bryan 49
Salter, Robert 265
Samsung Galaxy smartphone 170
Samsung Gear VR headset 168, 170
Samuda, Jacob and Joseph 255–9
Sanctuary AI - Phoenix 142
satnav 13, 124
Schwartz, Jacob 83
Science Museum, London 114
Science Research Council, UK 78
Scientific American 264, 265
Scott-Morgan, Peter 207–9
Scott, Travis 179
seasteading 331
Second World War 14, 73–4, 162, 224, 228
Sega 164–6, 180
Senning, Åke 190
Shakey robot, SRI 16, 118–26, 325
Shane, Janelle 89
Shannon, Claude 75, 76–7
Shockley, William B. 14
Shoulders, Kenneth 77
SHRDLU 78
Sidewalk Labs, Alphabet 307–13, 331
Simon, Herbert A. 77, 78
Sketchpad program 149, 150
smart cities 11, 13, 285–6
Alphabet Sidewalk Labs 307–13
Amaravati, India 314
Barcelona, Spain 317
BioDiverCity 314
Chengdu Future City, China 314–15
Clinton Foundation and Cisco Connected Urban Development Programme (CUD) 296–7
control rooms and dashboards 299–300
Cyberjaya, Malaysia 313–14
data collection and analysis 286, 290–2, 294–6, 319–20
Forest City, Malaysia 283–5
IBM 298–300
Kuala Lumpur, Malaysia 314
Los Angeles Community Analysis Bureau (CAB) 291–2
Masdar City, UAE 303–7
Medellín, Colombia 317
monitoring pollution 320, 321
NEOM and The Line, Saudi Arabia 315
New York City fires 1970s 293–6
privacy issues 309, 311–12, 318, 319
public transport 305–7, 318, 319, 321
Rio de Janeiro, Brazil 298–300
Santiago, Chile – Cybersyn project 300
Smart Cities Mission, India 316–17
Songdo New City, South Korea 300–3
Telosa new city design 314
Toronto Quayside and Waterfront plans 309–13
Toyota Woven City 314
traffic lights/management 287–90
Victorian London cholera epidemic 286–7
smartphones 9–10, 15, 96, 170, 172–3, 175–6, 193
Smolinski, Henry 230
Snap 176–7
Snow, Dr John 286
SoftBank 138
Solomonoff, Ray 77
Solzbacher, Florian 211
Son, Masayoshi 138
Sony 130, 137, 141–2
PlayStation VR 169–70
Soria y Mata, Arturo 315–16

space exploration 88, 119, 124, 162, 330–1
SpaceX 249, 250
Sproull, Bob 151, 153, 154–5
SRI International 15
Standford Research Institute (SRI) 15, 77, 118–22
Stanford Artificial Intelligence Lab 42, 45, 77, 91
Stanford Cart 126
Stanford Racing 42–3, 44–5
Stanford University 14, 63, 77, 82, 88, 126, 211
Star Wars, LucasFilm 156, 159
Stephenson, George 255
Stephenson, Neal 175
Stereoscopic Television Apparatus – Sensorama 151
StreetView, Google 45, 170
STRIPS program, SRI 122–3
Studebaker 224
Suleyman, Mustafa 94
Sullivan, Jesse 195–6
Sutherland, Ivan 128, 149–53, 171, 180, 332–3
Sutskever, Ilya 92, 98

Tagami, Katsuyoshi 134–6
Tampier, René 222
Tandy 15
Tange, Kenzo 316
Tatarinov, Vladimir 221
Taylor, Moulton 228–9
Technik Autonomer Systeme (TAS) 39
Teleological Society 86
Teller, Astro 323, 324, 330
Terman, Fred 14
Tesla 30–1, 47, 49, 62, 65, 141, 239
Thiel, Peter 98, 308, 331
Thrun, Sebastian 42–5, 47, 48, 50, 62, 172, 237, 245–6, 329, 330
Time magazine 305
Tonight Show Starring Johnny Carson 116–17
Toronto, Canada 309–13
Townsend, Anthony 309
Toyota 138, 314
Tracked Hovercraft project 266–9, 270
TRL GATEway pods 51–2
Trudeau, Justin 310
Trump, Donald 49
Tsukuba Mechanical Engineering Lab 39
Turing, Alan 10, 13, 66, 72–4, 86
Turing machine 85–6
Turing test 74

U-City and U-Life, South Korea 301–3
Uber 48, 49, 53–7, 64, 239, 242, 243
Unicorn Hybrid Black BCI 215
Unimation 116–18, 125
Uniroo, Leg Lab 128–9
Urban-Air-Port 242
Urban, Jack 187–8
Urmson, Chris 46, 49, 57, 60

Vallianatos, Mark 291–2
Valve - Steamsight VR headset 168
VaMoRs *(Versuchsfahrzeug für autonome Mobilität und Rechnersehen)* 32–6
VaMP and VITA-2 driverless sedans 35
vapourware 164–5
Velodyne 41, 47
venture capital (VC) firms 16, 94, 324, 325
The Verge website 176
Vertical Aerospace 237
Vidal, Eugene 223
Vidal, Jacques 210
View-Master toy 162
virtual reality (VR) *see* augmented reality (AR) and virtual reality (VR)
Virtual Visual Environment Display (VIVED), NASA 161–2
Virtuality 163–4
Vives, Antoni 300, 317
Volocity – Volocopter 236–7
von Neumann, John 14, 73, 75, 76, 86
VPL Research 158, 159, 161
Vuia, Trajan 220

Wainwright, Oliver 316
Wakamaru, Mitsubishi 137
Waldern, Jonathan 163–4
Walker, Giles 114
Wall Street Journal 134, 136, 139, 144
Warwick, Kevin 198–204
Washington Post 47–8
Watanabe, Katsuaki 138
Waterman, Waldo 223–4
Waymo project, Google/Alphabet 47–8, 49, 50–1, 54, 58, 59, 60, 61
Webster Rammell, Thomas 252, 260
Weiner, Norbert 74, 86
Weizenbaum, Joseph 78
Werbos, Paul 89
Wernicke, Ken 233
Westinghouse Electric Corporation – Televov 114
Whittaker, William 41, 42–3, 44–5
Widrow, Bernard 88
Wiegand, Daniel 234, 239, 244, 246–7
Wildcat, Boston Dynamics' 130
Williams, Ronald 89

Winograd, Terry 78
WIRED 10, 283, 313, 326
Wisk 237, 246
Wooldridge, Michael 69, 83
WordNet 91
World Wide Web 15
World's Fair, New York (1939) 20–1, 113–14, 332
Wright brothers 220
Wylie, Bianca 311, 331

xAI 65
Xerox PARC's Alto 15

Yokoi, Gunpei 166
York Commuter roadable aircraft 229
Yoshino, Hiroyuki 135
Young, James 197

Zeiler, Matthew 92
Zero Latency 156–7
Zimmerman, Tom 160–1
Zoll, Paul 190
Zuckerberg, Mark 79, 170, 178, 182, 210, 213
Zworykin, Vladimir 24